Civic Learning through Agricultural Improvement

*Bringing "the Loom and the Anvil
into Proximity with the Plow"*

A volume in
Studies in the History of Education
Karen L. Riley, *Series Editor*

Civic Learning through Agricultural Improvement

Bringing "the Loom and the Anvil into Proximity with the Plow"

Glenn P. Lauzon
Indiana University Northwest

INFORMATION AGE PUBLISHING, INC.
Charlotte, NC • www.infoagepub.com

Library of Congress Cataloging-in-Publication Data

Lauzon, Glenn P.
  Civic learning through agricultural improvement : bringing the loom and
the anvil into proximity with the plow / Glenn P. Lauzon.
     p. cm. – (Studies in the history of education)
  Includes bibliographical references.
  ISBN 978-1-61735-147-1 (pbk.) – ISBN 978-1-61735-148-8 (hardcover) –
ISBN 978-1-61735-149-5 (e-book)
1. Agricultural conservation–Middle West–19th century. 2. Service
learning–Middle West–19th century. 3. Social participation–Middle
West–19th century. I. Title.
  S604.62.M5L38 2010
  306.3'49–dc22

                                                          2010038043

Printed in the United States of America

Contents

Introduction

Agricultural Improvement as Civic Education

What do agricultural improvement and civic education have to do with each other? To a person who is familiar with the American Midwest's history, this book's subtitle triad of loom, anvil, and plow may be suggestive. To most people, though, the significance will not be apparent so readily. Discovering it involves three mental steps. The first step is to broaden our thinking about civic education, beyond participation in government, so that it incorporates what people actually learn about their community, its membership boundaries, and its allocation of obligations to different people. The second step is to suspend our inclination to associate learning with schooling, so as to become aware of education broadly conceived, to become attentive to the variety of institutions, experiences, and events that serve as important channels of educative efforts and socialization into community norms. With steps one and two accomplished, we can ask straightforward historical questions: How do people learn what is expected of good citizens? How do they acquire a sense of community?

The third step is to transport oneself into the nineteenth-century American Midwest. In that time and place most people farmed. The phrase, agricultural improvement, signified the aims and institutions that were intended to educate farmers to become better farmers. Peculiar as it may seem, men of the towns (some farmers, some not) were very involved in agricultural improvement. One of its principal aims was to grow towns. The loom and anvil of this book's subtitle symbolize the various manufacturing industries that townsmen hoped would take root as farmers learned

Civic Learning through Agricultural Improvement, pages ix–xviii
Copyright © 2011 by Information Age Publishing

ix

to farm—and to support the right kinds of public policies—in ways that bolstered economic development. In sum, the connection between civic education and agricultural improvement is that in the nineteenth-century American Midwest, better farmers were held to be better citizens because they lent their mite to the flourishing of civilization.

Few places in the Midwest could boast of having sizable manufacturing industries that brought income into a community; local farmers' support was crucial if such industries were to be planted and grown. In an age of relative isolation between places (the result of a lack of infrastructure and high shipping costs), townsmen wanted farmers to produce with an eye to selling and buying locally: supplying raw materials and foodstuffs for local use; marketing via local outlets farm products (especially wheat, corn, and hogs) destined for distant urban markets; and, purchasing from local retailers their farm implements, goods, and groceries. Farmers' economic decisions shaped greatly the scope and scale of the local economy. Residents of county seats were well aware of this. When they thought about the good community, about their community's future, they could not help but think about agriculture. Most staple crops and livestock were raised outside the incorporated limits, but how farmers farmed (and marketed) was a civic concern.

In this context, the civic problem was that few farmers considered themselves to be members of a community that included the county seat. They were farmers who lived in the country. Their attachment to place was restricted to their home farms, their sense of community obligation to neighboring farm families. Their income came, primarily, from supplying distant urban markets with grain and meat. Why should farmers care about a nearby town? What economic interest did they have in its prosperity and growth? The civic imperative of agricultural improvement was to alter farmers' working definition of community, to change their economic behavior so that it took into account—and created through multiplying commercial transactions—the interdependence of town and countryside.

Broadly speaking, the desirable pattern of farming behavior (and all that accompanied it) was referred to as scientific or progressive agriculture. To promote it, leading townsmen joined with leading farmers to create a host of educational institutions, including county fairs, agricultural societies (modeled on the scientific society and lyceum), the Grange (a voluntary association), and farmers' institutes (modeled after teachers' institutes). Aside from individual farmers' self-directed learning (through agricultural newspapers, conversations, and trial and error), these were agricultural improvement's principal agencies. Their efforts to bring farmers into the fold of agricultural improvement and economic development constituted a major source of Midwestern farmers' civic learning.

While I was doing the research for this book, numerous people asked me about it. My response was always to explain that I was investigating agricultural improvement for the purpose of revealing how agricultural associations and county fairs were a type of civic education for adult farmers in the nineteenth century. Often, a confused look passed over my listeners' faces. Part of the confusion originated in my claim to be an educational historian. Knowing that, people expected to hear something about the history of schools, schooling policy, or teachers' professional preparation. Another source of confusion stemmed from the fact that, formerly, I was a high school teacher of history and government. Knowing that, people expected to hear about the origins of the social studies or, at least, about how civic education was carried out in nineteenth-century schools. These expectations about the relevant sources and sites for the civic education of history formed a large part of my motivation for investigating nineteenth-century agricultural improvement.

What possibly could county fairs and agricultural associations have to do with teaching and learning in the school-based social studies? County fairs have featured schoolwork over the generations, certainly, and, farmers' organizations have influenced schooling policy and the ways schools carried out civic ambitions. This book, however, does not pursue these topics. It does not seek to reveal relationships between county fairs, farmers' organizations, and schooling in the nineteenth century. Nor does it try to establish connections between educational efforts among adult farmers in the nineteenth century and a brand of civic education for children that took root, initially, in large urban schools during the twentieth century.

Having declared what this book will not do, what it will do should be made explicit. The agenda and institutions of agricultural improvement will be explored as forms of civic education, forms that are both peculiar and appropriate to the nineteenth-century American Midwest. To achieve this, the book will focus on civic imperatives that leading farmers and townsmen insisted all farmers ought to heed. Sightlines will be constructed onto institutions of agricultural education (voluntary associations, fairs, and institutes) that were intended to make men better farmers and thereby better citizens. Within those sightlines, the objective will be to discern ingredients and processes of civic learning. To provide the rich description necessary to reveal the dynamics at work in the formation of civic consciousness within the rural community, this book will focus on educational institutions of agricultural improvement in Indiana.

Why a Case Study of Indiana?

To readers who are unfamiliar with Indiana, its selection for a case study may require explanation. Indiana has not received the kind of attention that historians, even those who study agriculture and rural areas, have devoted to other places. Proponents of progressive improvements often found mustering support to be a struggle in Indiana. In 1850, approximately one-half of Indiana's population had its origin in the South (principally from the upland region) and most of the state's people were farmers. As such, regional differences in conceptions of a good community and proper behavior co-mingled with the mistrust of innovations, particularly those associated with towns, that was characteristic of farmers throughout the United States. In connection with this, it is worth noting that in the nineteenth-century the words, civic and town, were used interchangeably as adjectives; among farmers, the connotation generally was not positive. This situation provides all the more reason to look for civic learning in Indiana. As a hard case of civic reform, it holds great potential for revealing the processes of learning that took shape as prominent townsmen and farmers tried to convince common farmers to change how they farmed and marketed.

For exploring the civic issues that intersected with agricultural improvement, Indiana's location and developmental stage are pivotal. In 1850, most of the United States was extraordinarily rural and undeveloped (in contrast to the long-settled and compactly populated parts of the northeastern states). Although it was admitted into the Union in 1816, Indiana was still very much like a frontier state. None of Indiana's 92 counties possessed more than one town of significant size, and only 24 counties had a town with more than 1,000 inhabitants. As settlers from the Northeast moved into Indiana, and as longtime residents acquired similar sensibilities, they expected Indiana's development to resemble that of the places they had left behind. In the Northeast, farmers were diversifying their market baskets to supply nearby towns and townsmen were encouraging manufacturing in all its forms. Indiana would have to follow suit, if it hoped to advance to the next stage of civilization. This overarching civics lesson about progress was something that Hoosiers would have to learn. It confronted people throughout the Midwest and, in subsequent generations, the Far West and South.

In the nineteenth century, much of the answer to the question, what kind of community should we live in? was supplied locally. In every state, a breadth of variety in geographic conditions and characteristics of the population demanded locally derived answers to such a fundamental question. State-mandated policy set (at best) broad parameters; the policy specifics—including the local option of choosing to adopt a policy or not—that gave shape and form to daily life were often decided at the county, municipal, and township levels. In matters of economic development, state legislatures tended to be permissive. They gave wide latitude to individuals who wanted to pursue new ventures in infrastructure-building or manufactures. Communities could support such ventures or not through referenda to raise special taxes or voluntary stock subscriptions. The approach was, in effect, the local option applied to economic development. It was an effective way to avoid disputes between communities in the realm of state governance. Adopting it meant, however, that there was not a single ideal for economic development or for what ingredients were preconditions for the good community.

Given the local orientation of the nineteenth century and the variety of conflicting influences—from southern old settlers and northeastern newcomers, from frontier conditions and civilized aspirations—within Indiana, it is striking that a rough consensus emerged on the fundamentals of economic development by the 1880s. By the close of that decade, the general tendency among Indiana's farmers was to support the civic package of agricultural improvement: self-education in diversified commercial agriculture, public policies that furnished a favorable context for manufacturing, and closer commercial ties between the countryside and its central towns. In comparison to the Indiana of the 1850s, this was a significant change. In their outlook and ambitions, Indiana's farmers had become much more like the inhabitants of towns; to a considerable degree, they had joined the civic community.

As they became more town-oriented, Indiana's farmers also became more like most of the people who lived in the rural communities of the Northeast and upper Midwest during the late nineteenth century. They recognized that a diversified mix of industries—agriculture, types of manufacturing, and various commercial enterprises—provided a stable foundation for community life. They placed a premium on self-improvement through education and effort in solid pursuits (not speculative ventures). And, they took pride in the progress their communities had made—in material and moral matters—when judged against the conditions of previous generations. People who lived in the rural areas of northern states expected that sort of progress to continue. But they did not expect or want life in their communities to become all that different from what it was at the moment (Baker, 1991; Barron, 1997).

The most southern of northern states and among the most western of eastern states, Indiana's people mirrored the attitudes of people throughout the rural North. For that reason, in the 1870s and 1880s, political party organizers looked to Indiana's early elections to see what fate November's elections would bring, and repeatedly chose Indiana's native sons as vice-presidential candidates. For that reason, in the 1920s, sociologists Robert and Helen Lynd chose Muncie, Indiana, as the "Middletown" of their ground-breaking study of American society (Lynd & Lynd, 1929). For much the same reason, Indiana serves as an excellent site for a case study of the civic dimensions of agricultural improvement in the latter half of the nineteenth century. As much as, if not more than, any other state, Indiana can stake a legitimate claim to being the representative American state of the late nineteenth and early twentieth century (Madison, 1986).

The Contours of the Case Study

For understanding the structure of this work, it should be kept in mind that its overarching purposes are, in a sense, conceptual and historiographical. The topical thesis is simple and straightforward: through the educative influences of and around agricultural improvement's institutions, the majority of Indiana's farmers became more civic-minded between the 1850s and 1880s. Showing the processes through which this tendency developed is the function of the story that is developed in Chapters 2 through 9. Book-ending the narrative are chapters that discuss conceptual and methodological issues. The first chapter sketches the ideas about civic education and historical research that govern the case study. As an extension of these ideas, it seeks to contextualize agricultural improvement within civic attitudes and practices of the nineteenth century. The last chapter is more explicitly concerned with historical method. Selecting aspects of the case study to use for illustration, it presents some considerations for investigating and reconstructing the civic education of history.

The case study's narrative begins with Chapter 2. That chapter establishes the civic and economic baseline for Indiana's promotional efforts in agricultural improvement during the 1850s. Of particular interest are the attitudes of Indiana's farmers toward economic development and the failure of a major attempt by the state to create a comprehensive system of transportation infrastructure. That failure left the state government with a massive debt burden and discouraged prospective migrants from the Northeast from relocating to Indiana. Determined to advance their communities' development prospects, Indiana's leading men sought to cultivate the ambition and initiative to improve among the citizenry. In other states, agricultural societies and fairs were demonstrating their effectiveness as inspirational educators. Indiana implemented legislation to create its own system of county agricultural societies and fairs in 1851.

Depicting how agricultural societies and fairs grew into high profile civic institutions during the 1850s is the purpose of Chapter 3. Organizers of agricultural societies and fairs learned quickly that most farmers would not be easily persuaded to join them. Prominent farmers and townsmen discovered new uses for the institutions. Gentlemen farmers found the county fair to be an excellent forum for displaying their wealth and status. Merchants found the county fair to be well suited for making money, particularly after railroad lines were built to connect their counties to the world. As these men took charge, the fairs took on new features, especially the horse race, that had little to do with the strictly agricultural mission to educate farmers.

On the eve of the Civil War, the agricultural fair had become a premier civic event, one that operated, primarily, on behalf of the county seat and its inhabitants. Chapter 4 depicts changes in fair-hosting during the years after the Civil War. Tendencies that had appeared in the late 1850s matured into their fullest expression: an agricultural fair that had little to do with farming and much to do with boosting the town's economy. Farmers were not pleased with the horse-racing, gambling, drinking, sideshows, and other attractions enjoyed by the town-dwelling crowds and young men of the countryside. That the scene of depravity and degeneracy was sponsored for the greater good of advancing agriculture (so the claim went) and economic growth made it all the more galling.

At no time were divided opinions between town and country on economic development more apparent than in the post-Civil War era. Sparked by the war and the pro-growth policies of Republican-controlled Congresses, the economy entered into its great industrial takeoff. Midwestern farmers were outraged at the impact on their lives. Mired in debt as the currency was contracted, crop prices sank, and prosperity appeared to bloom in villages, they reached a sensible conclusion. People in nearby towns (merchants and middlemen) and distant cities (railroad corporations and manufacturers) were reaping large fortunes at the farmer's expense. Farmers knew, too, that townsmen—both near and distant—were manipulating public policy to do it. Chapter 5 explores the farmers' political reaction, tracing the spread of a farmers' movement, the rise of a new organization, the Patrons of Husbandry (the Grange), and their impact on Indiana's politics during the election season of 1874.

Chapter 6 turns from politics and public policy to economics and activity in civil society. It reveals how farmers responded to the increasing control of townsmen over their economic affairs. Using the Grange's institutional structure, farmers engaged in direct cooperative purchasing to procure necessary goods, groceries, and implements from abroad (that is, outside the immediate county). Bypassing middlemen and contracting with distant suppliers, farmers circumvented townsmen's hold over local commerce. Deprived of farmers' dollars, middlemen and merchants felt the pinch of hard times in the 1870s. More important, though, for the long-term, the diversion of farmers' dollars abroad prevented their investment in local industries or in efforts to attract manufacturing. Aware of the harmful effects of farmers' economic cooperation on the town-based economy, leading grangers put an end to statewide cooperative purchasing, thereby depriving farmers from using the Grange as a weapon against the merchants and middlemen of nearby towns.

Prominent grangers joined other farmers in demanding broad reform in the political and economic system. On principle, though, the Grange leadership sought a moderate course. They would countenance only cooperative actions and public policies that gave due regard to the interests of the non-farmers of their communities. The Grange leadership insisted that farmers' true interests, properly understood, and the economic development of nearby towns were mutually supportive and beneficial. The Patrons of Husbandry, in other words, was not going to function as an exclusive interest group for farmers. It was a civic organization, one that claimed to be true to the nation's agricultural heritage, while working for the good of the community as a whole.

With some legitimacy, leading grangers could claim to be the spokesmen for farmers in the post-Civil War years. Certainly, they were better qualified than the members of county agricultural societies. Those organizations' main ambitions were to advance their towns' economic growth by holding big fairs. Given over to horse-racing and virtually any kind of feature that their directors thought might attract large crowds, county fairs had lost much of their agricultural character. Through the Grange, the curriculum of antebellum-era agricultural improvement was carried into the Gilded Age. Determined to purify the fairs of immoral tendencies and to give serious education in agricultural improvement pride of place on the program, grangers demanded reform. Chapter 7 examines farmers' objections and alternatives to the county fair of the 1870s, as well as their efforts to restore it to operating principles that reflected the moral and practical sensibilities of the rural community.

On the county fair's moral degeneracy, most farmers could agree quite readily. On political and economic matters, they could not. Differences of opinion nearly destroyed the Indiana State Grange in the 1870s. Most farmers wanted direct action against railroad corporations, merchants, and middlemen. The Grange leadership did not. Genuine grangers (that is, those who understood and subscribed to its doctrine) were too civic-minded. Their preferred method of reform was not to force town-dwellers to make significant changes in their economic behavior. Instead, grangers wanted to encourage farmers to study agricultural methods and economy so that they could adjust to the conditions of the post-Civil War economy. As farmers learned that their leaders were unwilling to place the farmer's self-interest over and above the claims of all other economic interests in their communities, they abandoned the Grange.

The Grange's declension left a leadership vacuum among Indiana's farmers that remained unfilled until the mid-1880s. By that time, as a result of hard experience and self-education, farmers were approaching a rough consensus on how they ought to respond to economic change. What they lacked was an institution to bring them together, to translate their personal experiences into policy, and to equip them with the practical knowledge needed by men who were willing to practice diversified commercial agriculture. Farmers in other parts of the country supplied a variety of programs and organizations as potential models. Ultimately, a broad coalition formed around a system of farmers' institutes that was jointly sponsored by Purdue University and the various farmers' organizations of Indiana's counties. The creation of that coalition in the middle of the 1880s is the subject of Chapter 8.

In its fundamentals, there was little that was new about the program of the farmers' institute. It envisioned profitable farms for farmers (through diversified commercial agriculture) and it placed those farms solidly within a local economic context of emerging manufacturing industries and growing towns. The agricultural societies of the 1850s had advocated more or less the same agenda. The civic mission of agricultural improvement had not changed; Indiana's farmers—and Indiana itself—had. The extent of that change became plainly evident in 1890, when a second farmers' movement swept into Indiana from the South. On the whole, Indiana's farmers were too intertwined with the economies of their towns for a crusade of the country against the town to attract their interest. The early part of Chapter 9 reveals the failure of agrarian populism in Indiana; the remainder explores how agricultural improvement's agenda and Hoosier farmers' civic learning may have dampened its appeal.

Chapter 10 is the final chapter. As noted earlier, it resumes the conceptual and methodological discussion initiated in Chapter 1. Between these bookends, to aid the reader in identifying civic implications, a set of reflections on civic learning is provided at the close of each chapter. In these brief reflections, the object will be to draw out the civic significance of what is described in the narrative. If it is successful, the mix of description and discussion will provide a persuasive explanation of how agricultural improvement helped to make Indiana's farmers more civic-minded. More important, this exploration of the civic dimensions of agricultural improvement and some of the learning derived from it, I hope, will offer insights to educational historians, civic educators, and others about the civic education of history.

References

Baker, P. (1991). *The moral frameworks of public life: Gender, politics, and the state in rural New York, 1870–1930.* New York: Oxford University Press.

Barron, H. (1997). *Mixed harvest: The second great transformation in the rural north, 1870–1930.* Chapel Hill: University of North Carolina Press.

Lynd, R. S. & Lynd, H. M. (1929). *Middletown: A study in contemporary American culture.* New York: Harcourt, Brace and Company.

Madison, J. H. (1986). *The Indiana way: A state history.* Bloomington and Indianapolis: Indiana University Press and Indiana Historical Society.

1

Locating the Civics in Nineteenth-Century Agriculture

Agricultural Improvement, Education, and Civism

This work seeks to redress two deficiencies in the historical scholarship of civic education. On the civic side, it highlights the role of farming in the well-being of the civic order during the nineteenth century. On the education side, it highlights non-school forms of education (adult voluntary associations and fairs) that contributed to rural people's understanding of their civic environment and their role in it. Through a topical focus on agricultural improvement, the case study reconstructs some of the processes through which civic consciousness formed among Midwestern farmers in the second half of the nineteenth century.

The relationship between agricultural improvement and civic education is not self-evident. Does not civic education refer to one's school-study of the United States Constitution and the means by which citizens participate in democracy? Does not agricultural improvement involve the promotion of new types of technology, crops, and livestock that help farmers make better use of their land and labor? There does not seem to be much of a connection. And there is not, for most people, in the United States today.

Civic Learning through Agricultural Improvement, pages 1–21
Copyright © 2011 by Information Age Publishing
All rights of reproduction in any form reserved.

In today's social environment, agricultural improvement involves primarily three groups of people. In experimental stations and laboratories, scientists engage in research that seeks to increase agricultural productivity and profitability. In rural communities, cooperative extension agents and agribusiness representatives promote the innovations in technology and technique that result from that research. On farms, those ideas and products are put to work by farmers and their employees in the business of raising crops and livestock. Another group of people can be classified as people who occasionally participate in agricultural improvement. When they purchase and consume food products, ordinary citizens are involved, especially when they consider the potential health risks and cost-saving benefits of, say, genetically modified wheat or pesticide use, that they have learned about via the newspaper, television news, advertisements, or conversation.

Along this continuum of activity, a great deal of education and learning occurs. Typically, few people are aware of it. Only the cooperative extension agent thinks of what he is doing as education, or, at least, as educational in nature. A scientist does research. The agribusiness representative thinks of his promotional activities as sales and service. The farmer who subscribes to farm journals thinks of his reading and consultations with experts as part of an ongoing search for ways to solve everyday business problems. The consumer, generally, does not think of his newspaper reading as an educational event. Keeping up with the news may be an important part of his daily routine, but it is not remarkable enough to warrant being designated as education. That word he is likely to reserve for the range of subjects associated with formal schooling, a curriculum in which agriculture has no place for most people.

In most cases, even when they are aware of themselves consciously acting as learners, few people in the United States today will perceive what they are learning about agricultural improvement as having any particular civic relevance. To be sure, sentimental images of farming remain a part of the American heritage, but farming is not central to most peoples' notions of the good life and the good society. It is only one sector of a highly complex economy, one that engages directly less than 2 percent of the United States population. Short of an environmental catastrophe that causes food scarcity or a food-based public health threat, few Americans will imagine civic implications in what farmers do. Given the distance between the dinner table and the farm, for practical purposes, the typical citizen's knowledge of modern farming needs to extend only slightly beyond the grocery-store aisles.

The diminution of farming as a pervasive presence in people's lives is one of the great changes in American society over the past century. Into the 1920s, the majority of the citizenry farmed or lived in small-town rural communities, where industries and commerce depended greatly on local agriculture. On the eve of the Civil War, more than 60 percent of Americans farmed and 85 percent lived in rural townships. In any state, city, or county of the United States it would have been difficult to find someone who was incapable of explaining, at length, the ways in which his livelihood was tied to the agricultural economy. In the sparsely populated and widely dispersed communities of the Midwest, it would have been impossible.

In the nineteenth century, a great swath of people's concerns about their communities revolved around farming and the impact of changes in it on the economic and civic order. Yet, on the whole, educational historians have failed to recognize the centrality of agriculture in nineteenth-century life and its bearing on educational affairs. Only occasionally does an educational historian give passing consideration to educational aims and institutions that targeted farmers (Cremin, 1980; Kett, 1994). We, today, are not farmers, so our first instinct in peering into the past is not to look for things related to farming.

Characteristics of contemporary society's approach to education are also responsible for drawing attention away from nineteenth-century agricultural improvement. Most people have spent enough time in schools to have learned to use the terms education and schooling interchangeably. Civic education, in particular, is equated with the school-based study of government. Prior to the implementation of custodial schooling (from ages 5 to 18) in the mid-twentieth century, these equations were not so fixed. People spent far less time in schools (three to five years' worth of short terms in the nineteenth century); school-based instruction, therefore, formed only a limited, but important, part of their education. The typical person's education—properly distinguished from formal schooling—stretched into adulthood in the context of events, public affairs, formal and informal institutions, and self-study. Educational historians are well aware that a great deal of education occurs beyond schoolhouse walls. Each year, the History of Education Society's call for conference papers "defines 'education' broadly, to include all institutions of socialization—mass media, voluntary organizations, and so on—as well as schools and universities" (History of Education Society, 2010). Nevertheless, after surveying the scholarly literature of the field recently, Milton Gaither (2003) concluded that "the history of education still means for most educational historians the history of school" (p. 161).

A vast terrain of educational institutions and experiences awaits investigation. Venturing into it, and assessing what we find there—how and what people seem to be learning—against what is happening in society and schools may be the most underappreciated methodological challenge in recent educational historiography. Reminding educational historians to look for educational efforts and learning wherever they take place is the historiographical purpose of this book. It carries out this purpose by exploring institutions that were intended to provide education in agricultural improvement to adult farmers living in the nineteenth-century Midwest. For the scholarship of educational history, the institutions and topic of this book are unusual; for being true to the experience of the nineteenth-century American Midwest, they are appropriate.

The men involved in promoting agricultural improvement did not refer to themselves as teachers; that designation was far too demeaning for men of their stature. They did, however, create educational institutions: annual agricultural fairs for broadcasting knowledge and inspiring farmers to pursue practical education; agricultural societies and associations for mutual instruction through reading, lectures, and discussion on a regular basis; and, agricultural newspapers for self-directed, home-based study. These educational strategies were not as intrusive or constant as those used in children's schooling; nor were they deployed in a distinct period of life that was devoted to preparation for adulthood. Designed for adults separated by geographic space, and grounded in voluntarism, they were based on the normal educational experience of the nineteenth-century, an intermittent series of episodes over the years that blended and moved back and forth between self-directed and peer-assisted study. These features serve to remind us that if we take into the past the same questions that we ask about today's schooling, we will miss much of the educational effort and learning.

In this book, a particular take on the educational historian's approach to the past is employed. It is inspired by the work of historians whose careers spanned the middle decades of the twentieth century (Bailyn, 1960; Cremin, 1965; Storr, 1961). As members of the *Committee on the Role of Education in American History*, these historians' gaze turned to channels of idea-diffusion and institutions of socialization, to the agencies that mediate between the ideas of society's elite and the experiences of the typical person. With publicity, funding, and their own works of scholarship, members of the Committee awakened temporarily other historians' interest in education broadly conceived and its potential impact on American society. Their approach—particularly as it was distilled and interpreted by Richard Storr, who wrote a memorandum for the Committee—provides the methodology for this work of historical reconstruction.

The Committee's approach might be readily grasped with a bit of simplification. Let us say that the general historian asks four primary questions about continuity and change: what? why? how? and with what effects? The educational historian is interested in these questions and something else, too: educational efforts in whatever guise they appear. That interest compels him to look closely in-between the usual benchmarks, reexamining webs of causality using three questions: (a) What ideas are expressed about how people ought to respond to conditions or changes in society? (b) What educational innovations are initiated? (c) How do people's experiences with those educational efforts feed back into the broader society? In this approach, the institutions and efforts through which (some) people attempt to propagate ideas take on special significance. The goal is to infer how individuals come into contact with ideas, take up those ideas (or not), and incorporate them into their behavior (or not). Few individuals leave a sufficient documentary trail of their educational experiences, however. As a result, the emphasis is on making inferences, using changes in the life of educational institutions as indirect evidence of what people were learning. (The approach and its implications are discussed at length in the concluding chapter.)

As a work of educational history, this case study of agricultural improvement serves as a vehicle for exploring how people's ideas about farmings role in promoting the good community were altered as a result of educational efforts deliberately set into motion. With its emphasis on community, the case study also draws upon the civic educator's perspective. On its most direct reading, the civic perspective attends to things that are of communal concern, as distinct from things that are of concern to a private individual or family (Noddings, 2006). In particular, the civic educator is interested in civism, defined generally as "the means used by society and/ or the state to cultivate the principles of the idealized citizen" (Dynneson, 2001, p. xiii). Two basic inquiry questions derive from civism. What are the behaviors and traits of the ideal citizen in a given community? How do people in positions of authority promote adherence to those traits among the community's membership? Schooling and education are two ways that the ideal citizen's traits are cultivated, but they are hardly the only means used to promote civism.

In this case study, the general or bare definition of civism will be used; no attempt will be made to craft a substantive definition that is peculiar to the nineteenth-century United States. The time span and extent of change involved, differences across the country's regions, and the variety of communities and perspectives defy any uniformly applicable and robust conception of civism. However, one can identify four ingredients that follow logically from, and give shape to, the general definition. These ingredients serve as useful markers for identifying what civism may look like and how it may operate in a particular time and place.

The first marker is the notion of place. What defines the geographic boundaries of the community? By virtue alone of living in a particular spot, people have (at least) common problems that result from living together. Recognition of a shared existence gives rise to the second marker, membership. Through political decisions, a corporate community that claims to represent the will—or the good—of the inhabitants must be created and given authority. Who belongs? What does it mean to belong? A collective sense of identity, attachment, or obligation must be fostered; on that basis, ideally, rests people's willingness to adhere to recognized behavioral norms that serve the well-being of the community.

The common, or public, good is the third marker. What is in the best interest of the community? How should people act to further it? Someone or some group of people must advance some notions; usually they are in leadership positions or possess high social status. Other people must be convinced to agree or, at least, to not interfere. The fourth ingredient is change. Invocations of what the public good demands almost always stem from actual changes or perceptions of impending change within the community. Reacting to change forms the rationale for mobilizing support for policies or prescriptions made on behalf of the public good. Why should people change their behavior? Civism supplies answers.

As a textbook-style rendering, this is too tidy. The parameters that define a place and the geographic boundaries of its community are often fuzzy. Who belongs to which place and why? The social boundaries of membership—its obligations and benefits—may not be perceived by, or conferred upon, all residents of a territory in the same way; inevitably, some members feel a greater or lesser attachment to the community and its ideals. Mechanisms intended to promote group identity may be too weak to foster a sense of shared fate. Problems that are identified by some people as common may not be recognized as such by others. The legitimacy of policies and prescriptions as to what the public good demands of people's behavior may be questioned, ignored, or openly defied. In the modern age, moreover, the infusion of aspects of democracy into the operations of communities opens the door to disputes over the nature of citizenship. Questions about who is included, and what the community can rightfully demand of particular members become embroiled in considerations of what ought to be done, why, and how. The civism of history is messy, not neat.

These complications and more figured into the civic ideals of the nineteenth-century United States. Civism, generally, was in flux. The decades that followed the American Revolution witnessed the waning of republicanism as a doctrine of leadership and the weakening of traditional notions of community authority. Simultaneously, the opening of American society to the opportunities presented by western land and an expanding commercial economy corroded place-based loyalties. Despite the deep impress place of origin made upon people's character, most people refused to stay put, moving from the farm into town, from one community to another within their home state, from one part of the country to another region entirely. Getting transient people to internalize community norms and to accept as authoritative the judgment of a community's leading men is difficult enough, but the difficulty was compounded by the rural nature of American life.

In origin, the ideal of a civic community is a product of the urban experience. In an urban center, population density creates a real and continuous need for behavioral consistency, if not a high degree of like-mindedness. From Ancient Rome and the Italian city-states of the early modern era, the Founders drew their civic inspiration. Among other issues, the condition of geographic dispersion plagued its application to the United States. Few Americans lived in cities or villages of a substantial size. Extending the limited degree of cohesion that existed in urban centers out to the countryside was difficult. As a result of the absence of high-quality interior roads, farm families were scattered and relatively isolated. Beyond the broad contours created by the commercial economy and the light imprint of government upon society, farm families' experience of the shared life was minimal and episodic. As a consequence (more often than not), to farmers, the civic community and its demands did not signify the public good, but what was good for the residents of the county seat or the nearest village.

Above all, what were justified in the name of the public good in nineteenth-century northern states were proposals to expand and diversify the local (town-based) economy through manufacturing, trades, and industries. Through its embrace of economic development, the town emerged as something distinct and separate from the countryside. As the town-dwelling life took shape, non-farming economic activities and the lifestyle features (dress, manners, refinement, and zeal for technological innovation) that accompanied them set townspeople apart from the countryside. To most farmers, these innovations were things of the town and of little concern to them. Convincing farmers otherwise was agricultural improvement's most basic civic mission.

Economic Change and Agricultural Civism in the Nineteenth Century

For much of the nineteenth century, conventional wisdom proclaimed farmers to be the ideal citizens for a republican form of governance. Near the end of the eighteenth century, in his response to Query XIX in the *Notes on the State of Virginia,* Thomas Jefferson provided the archetypal formulation of the connection between farming, the citizenry's virtue, and a republic's sustainability. Looking alone to God, Nature, and their own efforts, farmers—unlike merchants, tradesmen, and others dependent upon the "casualties and caprice of customers"—were as independent as any men could hope to be. Upon this foundation of economic independence, a virtuous citizenry could emerge, one in which each citizen managed his own affairs with prudence and the public's affairs with disinterest. Ancient history revealed the political lesson: the farmers' character and proportion in the population served as "a good-enough barometer" for taking stock of the "degree of corruption" in a society. With an immensity of land open to freehold farmers—and a growing European population to feed—the United States were uniquely favorable for sustaining a republican experiment (Jefferson, 1787/1984, pp. 290–291).

Thomas Jefferson's response to Query XIX looked in four directions for inspiration. Across the Atlantic Ocean, Jefferson saw the land "locked up against the cultivator" and the "mobs of great cities" employed in manufacturing. Around him, he saw an undeveloped countryside peopled with farmers whose homemade articles could not compete with the "finer manufactures" produced overseas. Ahead into the future, Jefferson saw a vision of (predominantly) agricultural communities immune from the social disorder that stemmed from concentrated wealth and contending economic interests. To the spirit of these farming communities, Jefferson looked to the ancient past, to the Roman Republic's ideal citizen: Virgil's *agricola*—a hard-working, virtuous, and independent farmer (Jefferson, 1787/1984, pp. 290–291; Wilson, 1981).

Thomas Jefferson's devotion to the farmer-citizen should not be confused with approbation for all farmers. He did not have in mind his half-literate hillbilly neighbors who barely scratched the soil when they tilled and let their hogs roam wild. Jefferson's praiseworthy husbandman was the small landed farmer (not planter) who was committed to farming and self-improvement. That kind of farmer was praiseworthy, in large measure, for his ability to distinguish needs from wants, and to keep wants in check out of concern for financial security. He would be no foe of economic progress, though. The good farmer would try, like Jefferson, to make his farm productive and profitable (Kennedy, 2003).

Ordinary farmers celebrated their economic independence and self-reliance. Their ideal was to use a combination of subsistence farming and commercial agriculture to secure a competency and comfortable subsistence, that is, to gain security in property and a reasonable standard of living (Vickers, 1990). Farmers might, as David Danbom (1995) wryly observed, "choose to be in the market a lot or a little, but they could not choose to be out of it" (p. 134). Even the most self-sufficient farmer needed a cash income to purchase his property, to pay taxes, and to buy some essentials, such as salt, utensils, knives, shot, and a few comforts for the home. All farmers were alert to the moneymaking potential of staple cash crops such as wheat, corn, hogs, tobacco, and cotton that could be sold in distant cities. For a variety of reasons, though, farmers in the rural North were learning to exploit other crops and livestock, things that, traditionally, were raised for family subsistence only. As they did, they began to recognize the advantages of having and supplying an agricultural market located in nearby towns and cities. Earlier and more avidly than Thomas Jefferson, they "learned to love manufacturing" as an essential counterpart to their farming (Peshkin, 2002).

A scene of rapid industrialization and urban growth dominates popular impressions of economic change in the northern states during the first half of the nineteenth century. Coastal New England, with its thin soil, short growing season, and rugged terrain felt these pressures keenly. But in most of the rural North—in upstate New York, central Pennsylvania, and the Ohio River Valley, for example—the growth of non-farming industries did not signal a rapid transformation in the nature of local life. Large cities were few and far between; market towns on waterways made little impression on the landscape. The stuff of daily experience made a deeper impression. With larger farms, better soils, and outlets to markets multiplying, northern agriculture flourished. As a consequence, people's expectation that farming would remain the predominant way of life held firm until after the Civil War.

Economic and social changes were penetrating the northern countryside and altering the nature of agriculture. They were not, however, eliminating farming as a viable way of life and occupation. From a broad historical perspective, therefore, the big questions facing farmers were more or less permanent. In what ways could commercially oriented farming be conducted so as to yield a comfortable subsistence while, relatively speaking, preserving a degree of independence? In what kind of community would they be farming as manufacturing industries and towns grew? What would be farmers' role and status? Should they remain farmers at all? With questions of this sort ever-present, agriculture and its prospects framed Americans' thinking about the good citizen and the good community throughout the nineteenth century.

In American society today, when people talk about good citizenship, political participation via voting is usually the first thing that comes to mind (community service often follows). In the nineteenth century, by contrast, participation in the formal operations of government was submerged in—if not eclipsed by—other forms of behavior that were associated with good citizenship. Politics for most people was something that occurred outside the routine of daily life; it did not require active participation on a regular basis. Indeed, political activity was often classified as something that was a corrupting influence on individual morality. The political parties' patronage system, scandals surrounding public officials, and electioneering debauchery (mass rallies, nonsensical slogans, vote-buying, whiskey-drinking, and election-day violence) provided ample evidence of that. Prominent statesmen might be honored for their public service but, for the typical person, it was far better to keep politics at arm's length, to attend to his farm, family, and fortune (Altschuler & Blumin, 2000; Saum, 1980).

Another contrast between the twenty-first and nineteenth centuries should be made explicit. In popular discourse today, the tendency is to take economic growth and most forms of development for granted, as things that are good for the community. The particulars of a proposal (for, say, a new industrial park or commercial center) may be disputed, but most people tend to agree on the premise that more is better, that economic activity ought to be encouraged so as to generate jobs and wealth in a community. This was not the case in the nineteenth century, particularly in the antebellum era. The tendency in public debates was to treat all economics-related policy proposals—tariff rates, taxation, and internal improvements, for instance—as political questions that had a bearing upon the way of life and nature of the community. The community at issue might be the nation as a whole, a specific region, state, or locality, but a policy proposal's potential impact on the community's character remained a key consideration.

Ultimately, all forms of economic activity carried significance for the community. In the rural North, this general interpretation of political economy created a civic context that tied the individual's behavior to the community's progress. A person's productive and commercial behaviors were evaluated in light of their contributions (or detriments) to the community's well-being. As character development and discipline, self-improvement in any honest calling or useful trade constituted the cornerstone. Efforts to improve one's condition were conducive to a family's self-reliance and benefitted the public well-being, particularly if one paid attention to relevant public policy. Regardless of a person's calling, the expectation was that he would engage in self-improvement and that his mite would be directed toward the good of the community. Far more than direct engagement in formal politics, virtues related to economic productivity became the criteria for gauging good citizenship in the nineteenth century (Berthoff, 1979).

In this context of citizenship, the much-trumpeted virtue of farming had little to do with formal politics and government. It had a great deal to do with the conditions necessary for creating the good citizen and community. At a minimum, farming protected moral integrity: the more attention devoted to the farm, the less time spent in the presence of town-based corrupting influences, such as taverns, amusements, gossip, and other vices. Additionally, farming done with intelligence and constant effort fostered the right kinds of dispositions, behavior, and character. A man improved himself through his farm-improvement. Indirectly (and when aggregated), the formation of personal character contributed to public well being: better farmer, better man/citizen. Likewise, the results of farm-improvement (when aggregated) shaped the pattern of life in local communities: improved farming, improved communities.

At the end of the eighteenth century, limited involvement in the commercial economy was the farmer's great virtue. As the bonds of the commercial economy strengthened and multiplied during the nineteenth century, most American farmers remained wary of becoming fully immersed in commerce, as producers and consumers. They subscribed to the early Jeffersonian notion that farmers ought to strive to be as independent as possible, that necessities for survival that could be derived from subsistence activities on the farm ought to be obtained that way. By setting a hard limit on their commercial wants, the need for cash income could be kept to a minimum. For cash income farmers relied, traditionally, on the seasonal sale of a single crop; they took pains not to become too dependent on the cash flow. In a risk-filled agricultural economy, commercial involvement carried a double-meaning: engaging in numerous transactions and obligations, as producers and consumers, and being in debt. Involvement of the first type often led directly to the second type. Debts often led to selling-out and relocating to a new territory or losing the farm. More involvement meant more risk, a greater threat to the farmer's independence.

To a far greater degree than, say, a town-dwelling craftsman, a farmer could pursue familial independence without becoming more involved in the commercial economy. As the decades passed, however, a farm family's reliance on subsistence agriculture marked its deviation from the emerging civic ideal of mutual interdependence. Explicitly promoted by the Jeffersonian-Republicans for a brief period straddling the War of 1812, the ideal of mutual interdependence was revived by Whigs as the harmony of interests in the 1830s. The call for fostering a harmony of interests was eclipsed by the rhetorically-more-appealing, if less revealing, ideal of progress in the 1850s. After the Civil War, for people living in the interior regions of the country, the ideal of progress coexisted uneasily with the call for home markets as the imperative. The catch phrases changed, the underlying meaning stayed the same. Good citizens contributed to the commercial economy; to the extent that they did, as members of the community, they would share in the ever-increasing standard of living and prosperity. If they refrained from full involvement in commercial activity, as producers and consumers, farmers were not considered to be good citizens.

This was the civic message, implicit and often explicit, of agricultural improvement. Acting as good citizens required fundamental alterations in farmers' private behavior as economic producers. Farmers needed to devote more energy to supplying nearby towns with a broad range of foodstuffs and raw materials. As farmers began to diversify their market baskets, they would gain more income—and more stable income—from frequent sales of different agricultural products throughout the year. From this, they would begin to discern how their economic interests were well-served by nearby towns' economic growth and they would support public policies related to economic development (railroad-building, local manufacturing, and road improvement, among other things). Through their economic interests in the town's commerce, farmers would become rooted and attached to the place-based community of town and country.

Almost all of these civic-economic prescriptions turned farmers' traditional wisdom upside-down. Yet, as civic moralists, agricultural improvers drew upon a full arsenal. They appealed to national patriotism and home pride; to Biblical injunctions and material gain; to family benefits and personal ambition, to public well-being and self-interest ("War and Politics," 1848). They ridiculed and shamed, drawing contrasts that seem humorous today but were scathing then, between farmers who preferred indolence to industry, ignorance to intelligence, and backwardness to progress (Snail, 1859). Above all, agricultural improvers demonstrated repeatedly that improved farming—done the right ways—paid and paid well (Cofman, 1854). Throughout, they tried to make the civic dimensions of agricultural improvement personal, that is, to equip the farmer with the understanding necessary to see his behavior as part of a much broader set of processes.

The failings of poor farmers aggregated into the ills plaguing society; improving farmers, in turn, lent their mite to social progress (Tuttle, 1857).

Advocates of agricultural improvement were well aware that more than the survival and comfort of a single farming family were affected by how a man farmed. They knew that advances in agriculture underwrote the emergence of civilization. Improving agriculture's productivity and diversification created the food surpluses, as well as many of the raw materials, that were the preconditions for the development of other industries and the flourishing of urban centers (Buel, 1838). It was not for nothing that variations on a remarkable expression penned by Jonathan Swift resonated in public life for nearly 200 years:

> Whoever could make two ears of corn, or two blades of grass, to grow upon a spot of ground where only one grew before, would deserve more of mankind, and do more service to his country, than the whole race of politicians put together. (Seldes, 1996, p. 443)

In public speeches at fairs, letters to the agricultural press, and editorials, men who never read or even heard of *Gulliver's Travels* (1726), hailed the man as a public benefactor who could make two blades of grass grow where only one grew before. At the nineteenth century's outset, the men who hoped to perform this miraculous feat were the rural community's leading men. By the century's midpoint, many farmers were taking the injunction seriously, particularly in the long-settled communities of the Northeast.

The Jeffersonian answer to European style urbanization ensured that improved agriculture remained central to nineteenth-century civism. As new communities were planted, they grew within a quasi-colonial relationship of economic dependency on the established East. What they needed, the East supplied, and they paid for it dearly. The creation of home markets—a diversified local economy and stronger ties to other places—was the key to breaking their communities' dependency (Fair, 2002; Roger, 1978). Even as leading citizens tried to attract manufacturing and other forms of economic enterprise to their towns, they tried to fill the countryside with farmers. To develop stable, complex, and economically healthy communities, they needed the right kind of farmer, the farmer who would contribute his mite to their communities' well-being through his productive activities.

Inescapably, since farming was the means of living for the majority of the American people in the nineteenth century, it was entwined in virtually every dimension of economic change. The processes involved—improved productivity, differentiation and specialization, and the integration of isolated places into a broader economy—are aspects of modernization. The social effects of modernization and their intersection with conventional ideals of republicanism have received a great deal of attention from historians in the past generation (Shankman, 2004). Investigating this interplay of economic, social, and political forces has caused historians to appreciate more fully the centrality of agriculture and its improvement in the mix of forces transforming American society. The fields of educational history and civic education have been slow to respond to this trend. In particular, the educational institutions devoted to agricultural improvement remain largely unknown. For this reason, it is necessary to summarize the development of agricultural societies and fairs prior to the 1850s, as necessary background to the Indiana case study.

A Survey of Agricultural Improvement in the Antebellum Era

Agricultural historians have described admirably modern agricultural improvement's origins. They can be found in England during the eighteenth-century, among learned scientific societies and within country gentlemen's efforts to improve their estates. The earliest agricultural societies in the New Republic were formed in the 1790s by men who copied the English gentry. Their conception of their public service role was circumscribed. It extended to improving their own farms (for profit-making potential and for others' emulation) and assuming the costs of conducting experimental trials of new crop varieties, livestock breeds, and implement designs. Serving as public benefactors in this way, the New Republic's leading men expected that, as ideas and innovations were determined sound, common farmers would adopt them. They did not, in any direct way, aspire to be educators of other farmers. Memberships of their agricultural societies were small and their activities limited to private meetings and personal correspondence. With a few exceptions, their duration was short-lived (Marti, 1979).

In the nineteenth century's second decade, another round of organizing took place. Encouraging farmers to improve their livestock, to practice soil-replenishing cultivation, and to broaden their range of commercially available produce were key aims. To these aims was added the goal of developing a self-sufficient economy, that is, a balanced economy of agriculture, manufacturing, and commerce. The desirability of economic self-sufficiency was underscored by the disruptions of trans-Atlantic shipping that surrounded the War of 1812. Nationalistic animosity toward Great Britain's manufactured goods spurred state financial-backing for the locally based agricultural societies. Wearing homespun, instead of manufactured cloth, was a symbol of patriotic solidarity.

To persuade farmers that a well-rounded economy was in their best interest, the first significant departure from the English learned society model emerged: the agricultural fair. One part competition for leading men, and one part education, entertainment, and festival for local families, the agricultural fair had the right ingredients to appeal to farmers. Adopted quickly everywhere, it got ordinary farmers involved in agricultural societies and became a prominent fixture of rural life. This, at least, is how the story was told by Elkanah Watson, the man who claimed to have the original idea of hosting agricultural fairs. Watson was not much of a farmer; he was, however, a man of civic zeal. If Watson amused his neighbors by venturing out to inspect a reputed potato-digging rooster, he surely annoyed them with his plugging for road improvements and canal schemes (Mastromarino, 2002). Watson was also a "remarkable publicist" and self-promoter (Marti, 1979, p. 15). Influenced in part by his autobiography, some historical accounts vastly overestimate agricultural societies' reach into the ranks of common farmers in the 1810s and 1820s (Kniffen, 1949; Savagian, 1998).

A major source of historians' confusion stems from the notion that a farmer is a farmer is a farmer. No one, however, would confuse Daniel Webster—owner of three farms and longtime United States Senator and presidential aspirant—with his Massachusetts neighbors, much less the typical frontier farmer (Sherman, 1979). The agricultural societies did little to aid future historians in making distinctions among nineteenth-century farmers. Seeking to reach out geographically into the countryside, and down socially into the ranks of common farmers, they asserted their members' credentials as plain farmers. Few were, in reality. Careful analyses of membership lists confirm what the political climate and educational standards of the age suggest. Agricultural societies were made up, largely, of professional men, town merchants, and country gentlemen. These men, like Elkanah Watson, were civic leaders who happened also to be farmers (Baatz, 1985; Mastromarino, 2002).

Given its civic leadership, the early agricultural fair did not function in quite the manner its agricultural mission might lead one to think. Having learned from observing the older generation that public service to agriculture was expected of the rural leadership, aspiring and ambitious men tried to use the fair's public competitions to bolster their social status (Thornton, 1989). Were not they acting as public benefactors? Were not they, too, worthy of deference and high standing? Some men were, surely, but giving awards for scientific essays, small-scale experiments with unusual crops, and imported livestock—all judged without regard to cost or profit potential—did little to persuade farmers that agricultural societies had a practical mission.

Opposition to agricultural societies mounted as fairs multiplied. The tide of popular sentiment was moving past inherited notions of republican deference by the 1820s. Farmers were making up their own minds about what was good for them and their communities. In New York, revisions to the state's constitution in 1821 brought into power a legislature more reflective of popular sentiment. Challenged to justify themselves, agricultural societies' defenders made their case. It was rejected. Bereft of public funds, all but one of New York's agricultural societies disbanded. The scenario was re-enacted everywhere by the late-1820s (Marti, 1979). As bearers of agricultural knowledge, the agricultural fairs and societies had little to offer; farmers had little appreciation for the civic knowledge they embodied (Gates, 1960).

The demise of the early agricultural societies and fairs does not mean that people lacked interest in improving their farms. The northern countryside witnessed a great deal of agricultural self-improvement. Considerable disagreement exists among historians as to why, exactly, and as to how the changes intertwined in rural social life. There is no disagreement on one point: farmers were farming differently. They cleared more land and put it under the plow. They experimented with a wide variety of grasses, grains, vegetable and fruit crops, and assorted species of livestock. They raised more of everything. Using their observations of other people's farms and farming habits, conversations at country stores and roadside fences, trial-and-error experimentation with innovative crops and techniques, and whatever information of value that could be gleaned from almanacs, newspapers, and the occasional agricultural journal that they might encounter, farmers were studying their farming. Hard work and the application of intelligence fueled the transformation in American agriculture that underwrote the industrial revolution. Adopting new technologies played almost no role in the change (McClelland, 1997).

The situation of agricultural societies changed in the mid-1840s. Bolstered by the example set by town-based lyceums and workingmen's associations, a new coalition of rural leadership formed to invigorate the pursuit of useful knowledge in the countryside. The coalition thought of itself as self-made, practically minded, and middle class. To a considerable degree, it was. Composed of working, but exceptional, farmers, town businessmen, and retired country gentlemen, this coalition blanketed the northern states with agricultural societies. Fairs once again were hosted widely. Public funding was restored. By 1860, every northern state had a State Agricultural Society or Board that served as the clearinghouse for county associations, educated the state legislature, and organized the latest innovation in agricultural education, the state fair. By 1858 a conservative count by the Commissioner of Patents identified 912 agricultural societies throughout the United States, three-fourths of which were in the North. Agricultural societies and fairs had entered their "golden age" (Neely, 1935/1967, p. 81).

What explains this dramatic change in agricultural societies' fortune? Four major factors are usually cited that, loosely fit together, form a thesis about the decline of northern agriculture. The factors are: diminishing soil fertility, westward migration, farm-leaving for town-based occupations, and the eroding social status of farmers. As farmers intensified their income-generating efforts in the early nineteenth century, the problem of soil replenishment spread. Working harder on worn-out soils, and getting lower profits on staple crops as their reward (courtesy of western competition), farmers gave up on their home farms and migrated westward. Or, they gave up on farming altogether to pursue more lucrative town-based occupations. As a way of life, farming was losing its appeal, the farmer-citizen ideal its luster, and the future of hundreds of rural communities were in jeopardy. This, more or less, is the thesis of agricultural decline in the northeastern states. The revival of interest in agricultural societies in the 1840s owes a great deal to the interplay of its main factors.

Agricultural improvement served as an alternative to soil exhaustion, declining farm income, and farm-leaving. As the population in settled areas grew, high quality land, particularly land well situated with respect to market access, was becoming scarce and costly. Forced to replenish their soil's fertility and to use available land more efficiently—and well aware that they could not produce wheat as cheaply as western farmers—northeastern farmers were focusing their efforts on dairying, vegetables and fruits, and cattle, hogs, and poultry to supply urban markets (Gates, 1960). To encourage the trend, some farmers took a lesson from the workingmen of the cities by organizing mutual improvement associations and promoting practically oriented agricultural newspapers. These papers served, effectively, as a monthly self-help service for farmers who were trying to adapt to emerging market conditions. Agricultural improvement's advocates were confident that farmers could continue farming in their home communities, if they learned to take advantage of advances in soil science, labor-saving machinery, and urban demand for foodstuffs (Marti, 1979).

The champions of agricultural improvement were correct. Given the drama of western migration and frontier settlement, it is easy to overlook the fact that many farmers resisted the broad acres of the West, stayed put, heeded agricultural improvement's prescriptions, and flourished (Barron, 1984). Despite the occasional historian's claim that agricultural societies and fairs were intended to "prevent agriculture from becoming 'lost' to the past" (Roston, 1993), there was no danger of agriculture disappearing. The conventional thesis of agricultural decline in the northeastern states, therefore, requires a qualification. Rather than declining, agriculture was changing its forms. In the process, it was sponsoring an emerging civic landscape, one characterized by small farms and small towns, stable economic growth, and a broad range of small-scale manufacturing enterprises, trades, and retail outlets. The balanced growth and interdependence of locally based economic sectors yielded increased efficiency, greater productivity, and prosperity for town and country. This formula and outcome represented the harmony of interests that political economists of the antebellum era idealized. Binding the countryside's farmers to central towns, agricultural improvement held their theory together (Conklin, 1980). Far from being backward-looking or an effort to prevent agricultural decline, it advanced the progress of rural communities.

By the 1840s, few people in the rural North pinned their hopes for their community's future exclusively on the condition of agriculture. A growing number did envision prosperity as contingent upon bringing town and country, manufacturing and agriculture, closer together. Using the agricultural press to build popular support, they called on state leaders to sponsor a network of county agricultural associations. Pivotal leaders needed little persuading, most notably Governors Allen Trimble of Ohio and Joseph Wright of Indiana. Hoping to lead the way in diversifying their states' economies, they became the first presidents of statewide agricultural associations. Equipped with leadership, public funds, and an institutional framework, agricultural improvement came into its own as an instrument for promoting the public good. Its advocates set their sights on three things: sustainable diversified agriculture, transportation outlets reaching out from interior towns to the broader world, and local manufacturing industries.

As inspiration for their ideal mix of development ingredients, agricultural improvers had the communities of central New York. In terms of good soil and tillable fields any number of places could claim to be a farmer's paradise. What made central New York different was location. Completion of the Erie Canal in the 1820s (followed by a host of feeder canals) gave New York's farmers what commercially oriented farmers everywhere wanted: ready access to multiple markets. As towns along the canal route multiplied, farmers within range could sell everything they raised with ease, if not at home, then, with the help of agents, to New York City and the world. New York's farmers made good on the marketing potential of their farmland's location. By the 1840s, localities across New York were famous for their cheese, butter, apples, and peaches (McMurry, 1995). Not coincidentally, New York's farmers were also famous for their unusual habit of reading agricultural improvement literature. In most states, farm papers died a quick death or eked out a marginal subsistence; in New York they thrived (Smith, 1971).

Urged on by readers of the agricultural press, New York State became the first state to create a network of agricultural societies around a State Board hub and the first state to host a State Fair. With its reading farmers and close commercial ties between town and country, New York's accomplishments pointed the way of the future, a future energized with the spirit of improvement and erected upon a broad economic foundation. By the late 1840s, leading citizens throughout the rural North were hurrying to apply the ingredients of New York's success at home. On the edge of the rural North and frontier West, Indiana's leading men joined them in earnest in 1851.

References

Altschuler, G. C., & Blumin, S. M. (2000). *Rude republic: Americans and their politics in the nineteenth century.* Princeton, NJ: Princeton University Press.

Baatz, S. (1985). *Venerate the plough: A history of the Philadelphia Society for Promoting Agriculture, 1785–1985.* Philadelphia, PA: Philadelphia Society for Promoting Agriculture.

Bailyn, B. (1960). *Education in the forming of American society.* New York: W. W. Norton.

Barron, H. (1984). *Those who stayed behind: Rural society in nineteenth-century New England.* New York: Cambridge University Press.

Berthoff, R. (1979). Independence and attachment, virtue and interest: From republican citizen to free enterpriser, 1787–1837. In R. Bushman, N. Harris, D. Rothman, B. Soloman, & S. Thernstrom (Eds.), *Uprooted Americans: Essays in honor of Oscar Handlin,* (pp. 97–124). Boston: Little, Brown and Co.

Buel, J. (1838). Matters of interest to all. *The (Albany) Cultivator, 3*(6), 89.

Cofman, S. (1854). More profitable farming. *Ohio Cultivator, 10*(3), 42.

Conklin, P. (1980). *Prophets of prosperity: America's first political economists.* Bloomington: Indiana University Press.

Cremin, L. (1965). *The wonderful world of Ellwood Patterson Cubberley.* New York: Teachers College.

Cremin, L. (1980). *American education: The national experience, 1783–1876.* New York: Harper and Row.

Danbom, D. (1995). *Born in the country: A history of rural America.* Baltimore, MD: The Johns Hopkins University Press.

Dynneson, T. L. (2001). *Civism: Cultivating citizenship in European history.* New York: Peter Lang.

Fair, J. D. (2002). The Georgia peach and the southern quest for commercial equity and independence, 1843–1861. *Georgia Historical Quarterly, 86*(3), 373–397.

Gaither, M. (2003). *American educational history revisited: A critique of progress.* New York: Teachers College Press.

Gates, P. W. (1960). *The farmer's age: Agriculture, 1815–1860.* New York: Holt, Rinehart and Winston.

History of Education Society. (2010). Call for Papers, History of Education Society Annual Meeting. Retrieved from: http://www.historyofeducation. org/2010/callforpapers.html

Jefferson, T. (1787/1984). Notes on the State of Virginia. In M. D. Peterson (Ed.), *Thomas Jefferson: Writings* (pp. 123–325). New York: The Library of America.

Kennedy, R. G. (2003). *Mr. Jefferson's lost cause.* New York: Oxford University Press.

Kett, J. (1994). *The pursuit of knowledge under difficulties: From self-improvement to adult education in America, 1750–1990.* Stanford, CA: Stanford University Press, 1994.

Kniffen, F. (1949). The American agricultural fair: The pattern. *Annals of the Association of American Geographers, 39*(4), 267–268.

Marti, D. (1979). *To improve the soil and the mind: Agricultural societies, journals, and schools in the northeastern states, 1791–186.* Ann Arbor, MI: University Microfilms International.

Mastromarino, M. (2002). Fair visions: *Elkanah Watson and the modern American agricultural fair* (Doctoral Dissertation). College of William and Mary, Williamsburg, VA.

McClelland, P. (1997). *Sowing modernity: America's first agricultural revolution.* Ithaca, NY: Cornell University Press.

McMurry, S. (1995). *Transforming rural life: Dairying families and agricultural change, 1820–1885.* Baltimore, MD: The Johns Hopkins University Press.

Neely, W. C. (1935/1967). *The agricultural fair.* New York: AMS Press Inc.

Noddings, N. (2006). *Critical lessons: What our schools should teach.* New York: Cambridge University Press.

Peshkin, L. A. (2002). How the republicans learned to love manufacturing: The first parties and the 'new economy.' *Journal of the Early Republic, 22*(2), 235–262.

Roger, W. W. (1978). 'The husbandman that laboureth must be first partaker of the fruits' (2 Timothy 2:6): Agricultural reform in antebellum Alabama. *Alabama Historical Quarterly, 40*(1–2), 37–50.

Roston, C. H. (1993). 'To make a better spirit': Community and history at the hill town fairs of western Massachusetts (Doctoral Dissertation) University of Pennsylvania, Philadelphia.

Saum, L. O. (1980). *The popular mood of pre-civil war America.* Westport, CT: Greenwood Press.

Savagian, J. C. (1998). *Whether to the haven or the maelstrom? The rise and development of agrarian reform doctrine in antebellum America* (Doctoral Dissertation). Marquette University, Milwaukee, WI.

Seldes, G. (1996). *The great thoughts.* New York: Ballantine Books.

Shankman, A. (2004). *Crucible of American democracy: The struggle to fuse egalitarianism and capitalism in Jeffersonian Pennsylvania.* Lawrence: University Press of Kansas.

Sherman, R. B. (1979). Daniel Webster, gentleman farmer. *Agricultural History, 53*(2), 475–487.

Smith, K. A. (1971). Moore's *Rural New Yorker:* A farm program for the 1850s. *Agricultural History, 45*(1), 39.

Snail, S. (1859). Farmer Snail on the *Ohio Cultivator. Ohio Cultivator, 15*(2), 20–21.

Storr, R. J. (1961). The education of history: Some impressions. *Harvard Educational Review, 31*(2), 124–135.

Thornton, T. P. (1989). *Cultivating gentlemen: The meaning of country life among the Boston elite, 1785–1860.* New Haven, CT: Yale University Press.

Tuttle, H. (1857). The tendency to spread. *Ohio Cultivator, 13*(22), 342.

War and politics—claims of agriculture upon government. (1848). *Ohio Cultivator, 4*(9), 70.

Wilson, D. L. (1981). The American *agricola:* Jefferson's agrarianism and the classical tradition. *The South Atlantic Quarterly, 80*(4), 339–354.

Vickers, D. (1990). Competency and competition: Economic culture in early-America. *William and Mary Quarterly, 48*(1), 3–29.

2

Between Frontier and Civilization

The Agricultural Improvement Agenda

Indiana's Farming Frontier

If any northern state needed an infusion of the spirit of agricultural improvement in 1850, it was Indiana. As part of the Old Northwest, the soils of Indiana had long felt the impress of white settlers. French outposts were dispersed at key points along the region's rivers in the eighteenth century. A rush of settlers and the installation of a territorial government under William Henry Harrison secured its status as a frontier state in 1816. Yet, despite the passing of 35 years and successive waves of an in-migration that swelled her population to nearly one million inhabitants, Indiana remained mired in frontier-like conditions.

The northern tier was a land of marshes, lakes, and streams. Thousands of miles of tile and open ditches would be needed to turn it into farms. In 1850, scarcely any drainage had been laid. To the south, in the central region, lay Indiana's greatest attraction for the farmer: a rolling plain, about 100 miles in extent, beckoned for plows. Axes were needed first. Indiana's best prospective farmland was heavily forested. Forests combined with swamps and hills to make the southern region of Indiana a daunting pros-

Civic Learning through Agricultural Improvement, pages 23–44
Copyright © 2011 by Information Age Publishing
All rights of reproduction in any form reserved.

pect for commercial agriculture. There the great ice sheet of prehistory left a patchwork of ravines and hills lodged between a series of rivers that ran due south to the Ohio River. Good farmland could be found easily enough, but access to the outside world was difficult. Indiana's geography was better suited to the recently evicted Indian tribes than it was to commercial farmers and enthusiasts for advancing civilization's progress.

Indiana had presented an almost perfect setting for its first two generations of white American settlers. Across the early decades of the nineteenth century they made their way into the southern part of the state. Leaving their homes in the hilly lands of North Carolina and Virginia, they traveled on foot through the Appalachian Mountains. Carrying what little they owned on their backs, on packhorses, and (if they enjoyed moderate prosperity) in carts, they took the Wilderness Road across Kentucky. Called "the longest, blackest, hardest road of pioneer days in America" by one authority, the overland trek was worth the hardship to the early Hoosiers (Philbrick, 1965, p. 306). Squeezed off their lands by the expansion of slavery-supported plantation agriculture, and faced with dwindling stocks of wild game and fertile soil, they found both in abundance. The new territory promised to be the closest approximation to the places they left behind that they could have found anywhere, a place where they could resume the familiar life they had always known.

The flat, treeless prairies of the West stirred the imaginations of a later generation, but the first Hoosiers preferred the hills. In the southern and eastern regions of Indiana, the land was naturally drained. Fresh water springs provided a ready supply of water for family and livestock; the absence of standing water kept families safe from malaria and fevers. The forest supplied wood for cooking, timber for building (and sale), fowl and game of all kinds for eating, and space for hogs to roam. Starting with a small clearing for a cabin, garden, and cornfield, a family could grow a farm, adding a few acres of cultivated land each year by girdling and felling trees, lightly plowing with a one-horse jumper plow, and planting corn among the stumps until they rotted sufficiently to be uprooted. In the meantime, hogs could live off the nuts, berries, and plants in the woods until it was time to herd them into pens for fattening on corn prior to driving them overland to market (Cayton, 1996).

Living in loosely grouped settlements, these Hoosiers carried on a lifestyle that historian Frederick Jackson Turner (1950) characterized as "rural separatism" (p. 267). Coming together occasionally for events that blended serious work with frivolity and socializing (husking bees and log-rollings, for example) families spent most days on their farms. The daily routine of frontier life was interrupted by neighbors and travelers passing through;

the seasons broken up by trips to the county seat some miles away to purchase supplies and, if they chose their day shrewdly, to be entertained at
the courthouse by country lawyers arguing cases before a traveling judge. A
few tools, luxury food items and household comforts (metal blades, coffee,
sugar, silverware, shoes, for example), as well as the occasional service from
a town-based tradesman, were the only necessities the farm and forest could
not supply. Occasional outwork, seasonal employment in leather-tanning
yards and pork-packing houses, and sales of forest products supplied some
of the cash needed to pay taxes and to purchase these essentials. Annual
hog sales, however, represented Hoosier farmers' most important entry
point into the modern economy (Carter, 1946; Esarey, 1917).

Operating on the margin of survival at the commercial economy's
fringe, aside from hog sales, Hoosier settlers did not expect their farmland
to generate much in the way of profits. Instead, they pursued the strategy
that agricultural historians call safety first: since most needs could be met
on the farm, they purchased from the market selectively and produced for
it opportunistically. Most farmers were reluctant to improve their farms beyond a certain point. On the frontier, improved livestock, implements and
machinery, and extra labor were scarce and expensive. Making forest into
farmland was a sure thing: steadily increasing settlement virtually guaranteed a profitable resale of the farm. Making cleared farmland into highly
productive farmland required numerous small investments that might jeopardize a family's immediate financial security. Thus, the precariousness of
the agricultural economy reinforced attitudes acquired over generations of
semi-subsistence survival in the upland South. Together, these two factors
conspired to make converting Indiana's wilderness into a flourishing rural
society a slow process (Nation, 2005).

Outside observers, generally, were not convinced that a majority of
Hoosiers wanted to hasten their state's development. Hoosier farmers had
acquired an infamous reputation for hostility toward Yankee notions of
progress and improvement. They were often characterized by New England
migrants and foreign travelers as shiftless farmers who preferred to hunt
and fish, and to lounge in the shade with a whiskey bottle, swapping tall
tales and political gossip. Grounded in chauvinism and a rejection of the
rationality of frontier life's alternating periods of strenuous work and between times, the caricatures were surely overdrawn. Nevertheless, Hoosier
farmers displayed less enthusiasm for improving their farms than respectable society insisted they ought (Power, 1953).

Hoosier deviance displayed itself in Indiana's economic geography.
Measured against the settled East, Indiana's settlers had accomplished

little. To be sure, they had removed the Indians. What the early settlers had not done was more striking. Only 5 million of the state's 23 million acres were improved for pasturage and cultivation. Rather than fencing-in their livestock, Hoosier farmers fenced them out: they fenced in their crops instead. Waterways, more so than roads, connected the few towns of significant size. The Wabash River (with a canal extension) connected Fort Wayne, Lafayette, and Terra Haute; with the exception of Indianapolis, the remaining cities were strung along the Ohio River. Only two major roads existed, the National Road and Michigan Road, bisecting the state's center on the east-west and north-south axes. Even these roads were rough affairs. Stumps remained to break wagon axles when the roads were muddied, and logs were laid crossways across wet spots. Local roads were little more than wagon tracks. Away from the main rivers and the two prominent roads, once a traveler left a town's immediate vicinity he moved through a landscape that differed little from when it was occupied by tribal peoples (Barnhart & Carmony, 1954).

As they interpreted it, at least, Hoosier farmers had taken Thomas Jefferson's early vision of a decentralized society of independent farmers to heart. Their vision, however, did not match Thomas Jefferson's mature vision of placing the farmer, manufacturer, and merchant side by side in rural development. Nor did it match the vision advocated by Whig and Democratic leaders who claimed the Jeffersonian mantle in the 1840s and 1850s. What community leaders had in mind was a complex of largely self-sufficient villages surrounded by well-maintained farms and connected to other towns through a developed road system. Jefferson's political heirs did not have as their ideal a scaled-down version of the southern plantation pattern. Most Hoosier farmers did.

The rough-hewn frontiersman has come to dominate modern images of the Old Northwest's settlement. Interspersed between the landless squatters and upcountry farmers of the pioneering generation, however, were men of substance and enterprise. They did not emigrate simply to get a living. The cabins they erected were replaced quickly by brick and frame houses of ample proportions. Those who wanted to farm did so with zeal, adding field to field, planting orchards, and going to great lengths to import livestock worthy of tending. Others preferred not to farm, but had to, at least, on a part-time basis. Equipped with the trade skills and book-learning needed, they set up sawmills and gristmills, dry goods stores, brick-making operations, and other small-scale enterprises. They came to frontier Indiana to make lives and communities that rivaled—or surpassed—what they had left behind. Abreast of what was happening in other states, they urged their fellow Hoosiers to support measures to create stronger commercial ties connections with the Atlantic seaboard (Madison, 1986).

The Mammoth Improvements Project

To the improvement-minded, Indiana's greatest obstacle was its natural terrain. With no major rivers running through the state's interior, access routes needed to be opened and improved. When Indiana was admitted into the Union, older states were speedily building networks of turnpikes connecting river cities to inland towns, and smaller roads linking towns to the countryside's farms. A great wave of canal building followed. The improvement-minded were acutely aware that there were alternatives to Indiana's grueling travel conditions. Virtually "[e]very message of every governor discussed internal improvements, frequently in great detail" (Barnhart & Carmony, 1954, p. 276). Newspaper editors championed the cause, lauding successful transportation projects in the older states. A system of roads and canals was needed, one that brought together the densely populated southern region with the scarcely settled northern tier, one that connected the Whitewater river valley in the east with the Wabash in the west, one that reached out from Indianapolis to every border and every isolated rural pocket.

Blanketing the state with a system of internal improvements proved difficult. Cash-poor (like most citizens), Indiana's government was forced to rely on its congressionally allocated percentage of public land sale proceeds. In 1821, the General Assembly appropriated $100,000 for building a road system; land sales yielded $40,000. Distributed across Indiana, it was not enough to make lasting improvements to existing roads or to launch major projects. Leaders representing communities that would have been bypassed in the routes of proposed projects refused to give up their share. Into the 1830s, frustrated governors urged members of the General Assembly to set aside local interests for the good of the state. Their pleas went unheeded. By 1837, the state had little to show for the $430,000 that had been distributed.

The exception to Indiana's dismal internal improvement record was a striking one. Construction of the Wabash and Erie Canal began in 1832. Joining the Wabash and Maumee rivers, it proved an immediate boon to Lafayette, Fort Wayne, and the counties along its route. The construction itself sparked a boom. With large numbers of laborers to feed, prices for agricultural produce soared. Access to eastern markets boosted prices further still, while distance readily spanned gave farmers great savings in transportation cost, time, and aggravation. As emigrants poured in and awareness of the canal's benefits penetrated the state, confidence gathered momentum. What the public land sale revenues could not finance, banking on the future—using the public credit—might. New York, Pennsylvania, and Ohio had used public debt to improve their states. With the immediate success of the Wabash and Erie Canal before him, in 1834, Indiana's governor thought—and many people agreed—that there was "no good reason" why Indiana "should longer hesitate to follow" their example (Barnhart & Carmony, 1954, p. 297).

It took two years to reconcile competing local interests, to craft a package of projects that would secure sufficient public support throughout Indiana's regions. The resulting proposal offered something to (almost) everybody everywhere. Three major canal projects were planned. On the state's western side, the Wabash and Erie Canal would be extended south to Terre Haute. Another, the Central Canal, would branch off the Wabash and Erie's northwestern section to connect to Indianapolis, and then extend all the way to the state's southern border at Evansville. The third, the Whitewater Canal, would open the central region to the east, by linking the National Road to the Ohio River. Beyond the canals, several other projects were envisioned. The southern interior would get a macadamized road running from New Albany on the east to Vincennes on the west. A railroad line (the first of significance) would cut through the state on a southeast-northwest diagonal, joining Madison to Indianapolis and Lafayette. Surveying for future routes for canals, roads, or railroad lines would follow.

The general improvements proposal, commonly known as the Mammoth improvements bill, carried a whopping price tag of ten million dollars. Borrowed through bond sales at 5 percent interest, annual interest payments alone would equal $500,000. At the time, the state's annual revenues amounted to only $75,000, but supporters of the plan were confident that the projects would pay their own way as the communities along the envisioned routes grew. It was a reasonable expectation, broadly shared. Much of the grumbling that occurred among the public was not about the Mammoth proposal's scale or riskiness; it was, instead, complaining that more projects should have been added. After all, the Erie Canal had transformed New York's interior, as the Wabash and Erie was doing in Indiana's northern tier. With Indiana's population already doubling each decade, the drawing power for new migrants, as well as the economic activity the new projects would bring, could furnish a rate of growth that would make the state's debt appear to be a modest investment in the future.

Everything was right about the Mammoth proposal except the timing. A financial panic swept the nation in 1837. The depression that followed was prolonged, particularly in the agricultural export dependent states of the Old Northwest. Work on the Mammoth projects continued until funds dried up in 1839. Only 90 miles of the Wabash and Erie extension had been built, 30 on the Whitewater Canal, and 28 on the Madison-Indianapolis Railroad. Hundreds of miles of anticipated projects around the state lay in varying stages of completion. The projects remained stalled into the 1840s, as representatives of the state negotiated with financiers on the loan repayment schedule and half- million dollars in interest that were piling up each year. The state government was mired in debt until well after the Civil War (Barnhart & Carmony, 1954).

The Mammoth improvements plan's failure dealt Indiana a crippling blow. While work on the projects languished, migrants from the East bypassed Indiana. In the 1820s and 1830s the state led the way in attracting new migrants to the Old Northwest; in the 1840s, in-migration slowed to a trickle. Stamped by the first wave of northeastern migrants as a land of malaria and bogs, southerners and hogs—and now plagued by rumors of impending tax increases—Indiana held little appeal. Easier traveling and unsettled country drew migrants to Michigan and Wisconsin, while prairies ready for the plow lured the land-hungry to Illinois and beyond (Cayton, 1996; Power, 1935).

Some potential emigrants may have thought that Indiana's lands were full. Counties along the Ohio River rivaled the rural communities of the Northeast in population density. Indiana's cultivable lands were far from overstocked, though. The population density of much of the state's interior region barely surpassed that of newly opened western lands (Rose, 1985). Native Hoosiers were advancing steadily into these areas. For settlers to move from river valleys to less accessible locations over time was normal. However, it could not be denied that growing numbers of Hoosier sons and daughters were leaving the state. The reason often given for their departure may have deterred potential emigrants from moving to Indiana.

Reports of soil exhaustion in Indiana were becoming common by the 1840s and 1850s. As William Dennis, a leading agricultural improver of Wayne County, informed readers of *The* (Albany) *Cultivator*, the nation's most prominent agricultural paper, "many of the original settlers" were "the finest exemplifications of the *'skinning system'*": "having cropped with corn after corn" until it was questionable whether the harvest yield "would equal the seed," they sold out and moved on (Dennis, 1850, p. 369). Dennis's intention was to highlight the advances being made in his home county. He described at length the opportunities Wayne County presented for migrants and took pains to stress the land's capacity for restoration under energetic ownership. But, as the reports of soil exhaustion circulated through eastern communities, readers and hearers unfamiliar with Indiana's geography and early settlement likely concluded that the state did not match the prospects elsewhere. Few farmers who were accustomed to relatively easy access to markets in the Northeast, in any event, were willing to purchase even the most fertile interior lands without the assurance that improved infrastructure was being constructed.

Attracting new immigrants and investment through vigorous publicity efforts and high profile infrastructure projects was how frontier state boosters typically tried to grow their states. By 1850, it was clear that the strategy had not—and would not—work in Indiana. Historians have suggested that the Mammoth plan's failure engendered a sense of resignation that the state could never catch up with others (Abbot, 1978; Thornbrough, 1965). However, with a heavy public debt and the public's mistrust at an all-time high, there was little that the state government could do to promote economic development. For progress to occur, the initiative would have to be cultivated among Indiana's citizenry and the state's development hopes anchored firmly on their self-improvement ambitions. To a considerable degree, that solution corresponded to traditional notions about how communities should grow, that is, gradually, through natural increase, not rapidly through infusions of people and capital. It also matched the experiences of men who came of age in the initial decades of Indiana's statehood. As the Mammoth improvements projects foundered, these men were assuming prominent roles in public life and looking to use agricultural societies and fairs as means of advancing their communities' fortunes.

The Improvers: A Profile

Indiana's leading men had life experiences similar to those of other, more famous, men of her sister states. Members of the first generation born after the formation of the New Republic, these new men of the rising West—men like Thomas Hart Benton of Missouri and Stephen A. Douglas of Illinois (and, of course, Abraham Lincoln)—were acutely aware of their place in history. Transplanted to the Old Northwest at an early age with their families, or setting out on their own at the age of majority, they came, often, from modest circumstances. Provided the rudiments of learning by their families, they pieced together the learning they thought would be necessary to their success in life as they encountered exceptional older men who served as their mentors and models for emulation. Driven by ambition, they worked as farm laborers, teachers, carpenters, and clerks while searching out opportunities to advance. They saw in their personal lives, writ large, what the future held for their states. Whether from Illinois, Missouri, or Indiana, they envisioned their state growing from a pioneering backwater into a leading state of the Union (Turner, 1950).

Biographical information on the men who took the lead in promoting agricultural improvement in Indiana is sparse. Practically minded men who moved several times in their lives, they left few records for posterity. Compilations and county histories produced in the late nineteenth century, however, provide sketchy information about some of them. The men who were profiled were not typical Hoosiers. Rather, they were the kind of people they (and others elsewhere) thought would determine the future (Cayton, 1996). Their character type is well known, so only a few examples are necessary to illustrate the men and motives behind the launching of agricultural improvement in 1850s Indiana.

One of the few men able to act on a statewide basis, Governor Joseph A. Wright was a driving force in Indiana's agricultural improvement campaign during the early 1850s. Approached by two State Senators, one who chaired the Agricultural Committee and another who edited Wayne County's leading newspaper (and would soon launch an agricultural paper, the *Indiana Farmer*), Governor Wright took up the cause with enthusiasm. An earlier state law to promote agricultural improvement, passed prior to the financial panic of 1837, had left few lasting results. Lacking leadership, and impeded by the traveling conditions of frontier society, few agricultural societies were formed. Less than 10 were identified (existing, largely, in name only) by the Senate's Agricultural Committee in 1847 (Carmony, 1998). A staunch advocate of economic development, but no fan of state-administered programs, Democratic Governor Wright seized the opportunity to bring agricultural improvement before the state legislature and the Hoosier people.

Little in Joseph Wright's background suggested that, at the peak of his life's ambitions, he would become a promoter of agriculture; virtually everything indicated that improvement would gain his avid support. Born in Pennsylvania in 1810, he moved to Bloomington, Indiana, with his family at age nine. Obtaining some schooling between time spent at work with his father and brothers in a brickyard, he was among the few people who were at the right place at the right time to acquire some higher learning. Soon after his father's death in 1823, he enrolled in Bloomington's newly established college (later Indiana University). Earning his keep by performing janitorial work, working odd jobs in masonry, carpentry, and implement repair, he left college after two years to study law with a practicing attorney. His apprenticeship over, he struggled to establish a law practice in Parke County, while starting a family with his new wife and teaching Sunday school for the

Methodist Church. Denied a postal contract in 1830 because he "was not sufficiently known or recommended," he entered local politics and gained election to the state legislature three years later (*Eminent and Self-made Men,* 1880, pp. 247–248). A career in politics followed, with terms of service in the General Assembly and United States House of Representatives interspersed with periods of private practice. A hotly contested campaign yielded him the governorship in 1849 (*Biographical Directory,* 1980).

Serving as President of the State Board of Agriculture, Governor Wright stumped the state widely, lecturing all who would listen on the advantages of farmers' associations and urging farmers to experiment with new crops and improved methods. Wright had never been a farmer. His agricultural addresses mixed conventional Jeffersonian doctrines of self-reliance with advocacy for diversifying the state's economy. For his agricultural content, he relied on stock phrases and anecdotes gleaned from farm papers. He called on farmers to make two blades of grass grow where none had grown before and told hackneyed stories of slipshod farmers, like the one about the man who built his barn on a hillside so the rains could wash away the manure. Wright's political opponents took advantage of his new interest in agriculture. Repeating a tale that was being used throughout the country to lampoon agriculturally ignorant politicians (that the political figure suggested farmers improve their sheep by purchasing hydraulic rams for breeding), they charged Wright with currying favor among the farmers (*Eminent and Self-made Men,* 1880). A novice to agricultural improvement, Joseph Wright certainly was, but he was sound on the needs of Indiana's economy. He took his duties as President of the State Board of Agriculture seriously, attending lectures, participating in discussions, and mustering support for the cause.

Governor Wright may have been open to the charge of political opportunism, but no one could accuse Joseph Orr of using agricultural societies for personal advantage. As Wright had been, Orr was born in Pennsylvania. Among the first settlers to move into the Old Northwest, at the age of five he found himself living the life of a squatter a dozen miles to the north of Cincinnati (population around 600) at the turn of the century. A boyhood of tree-chopping, root-grubbing, brush-burning, and pioneer farming followed. With three scattered terms of schooling under his belt (and ten other children for his father to support), he was apprenticed to learn carpentry. Marriage to a Yankee schoolmarm and a few years as an independent carpenter followed. Several trips to Indiana, in the meantime, convinced him of its favorable prospects. In 1823 he relocated his family and began making a farm near Greencastle (*Eminent and Self-made Men,* 1880).

Clearing farmland and selling dry goods on the side, Joseph Orr rapidly made himself a part of the local community. His business, combined with service in the Indiana militia and acting as a manager of the local bank, secured his local reputation. Nominated for the General Assembly, he served as a state representative and senator. When prairie farmlands near Lake Michigan beckoned, Orr's career in public office ended. Leaving Greencastle in 1833 with his family and farmhands in a procession of hogs, cattle, and wagons, he set about turning a 1100 acre tract into farmland. A successful farmer and rancher (with saw milling and wood-cording operations), Orr acquired extensive holdings, before parceling-out sections to his sons as they came of age. Nearing the end of his active period of life when the State Board of Agriculture was chartered in 1851, Joseph Orr had nothing to prove or gain from promoting agricultural improvement. Widely known for his militia leadership during the Black Hawk War, General Joseph Orr had a reputation that carried weight in public opinion; he had the leisure to frequent agricultural meetings and fairs; and, most important, he had credibility with farmers (Daniels, 1904).

Similar to Joseph Orr, another charter member of the State Board of Agriculture, Alexander Campbell Stevenson, enlisted in agricultural improvement as a retired gentleman farmer. In contrast to Orr, who seems not to have engaged in other public improvement efforts, Stevenson was a tireless champion of the three most famous reform campaigns of the antebellum era. Born in Kentucky in 1802, Stevenson in his youth acquired (contrary to his father's practice) an aversion to slavery. Living in Indiana at the time of his father's death, he brought the family's slaves to Indiana and freed them. Among the first temperance lecturers of Indiana, Stevenson's confidence in "moral suasion" seemed unaffected by the men who "drank bumpers to each other and to the speaker" as he decried the "evils of intemperance" (Weik, 1910, p. 698). A college-trained physician, Dr. Stevenson urged common schooling for all, and higher education for those who sought to obtain it. With almost twenty years of service in the General Assembly, Stevenson was well-positioned to advance these causes until he abandoned public office for good out of disgust with Indiana's constitutional convention of 1850–51, to which he had been a delegate.

Unlike Governor Wright, Dr. A. C. Stevenson did not need the formation of the State Board of Agriculture to inspire his interest in agricultural improvement. From the time he retired from his medical practice in 1843 (at the age of 41), it occupied his attention. During his years as a country doctor he had started making a farm two miles east of Greencastle. Acres of timber had been replaced with Kentucky blue grass well suited for pasturing thoroughbred horses and cattle. He took the lead in importing these into

Indiana through the 1840s and 1850s (through self-financed trips to England, Kentucky, and elsewhere). Hoping to encourage others to do likewise, he tried to organize a joint stock company for the purpose. Devoting some of his energies to writing for the weekly agricultural column of a county newspaper, he was widely known and respected as a leading farmer in central Indiana. Living to the ripe age of 87, he joined a much younger set of farmers in establishing the Patrons of Husbandry (Grange) in Indiana and organizing a state wide Shorthorn Breeders' Association. His golden years were filled with days spent advising farmers at county fairs and grange meetings on how to improve their farming and how best to generate political support for their cause (Weik, 1910).

In advancing agricultural improvement, Joseph Wright, Joseph Orr, and Dr. A. C. Stevenson were joined by a host of leading men from counties across Indiana. Some, like James D. Williams, who was elected to serve as Indiana's Governor in the 1870s, were lifelong farmers who played prominent roles in local and state politics (Shoemaker, 1981). Others, like Thomas Kirby of Delaware County, avoided public office. They focused on growing their fortunes by expanding their farms and selling thoroughbred livestock, and, in Kirby's case, engaging in land speculation (Helm, *History of Delaware County*, 1881; *A Portrait and Biographic Record of Delaware County*, 1894). Some, like James Cockrum of Gibson County and I. D. G. Nelson of Allen County, split their time between farming and other lines of enterprise. Through his lifetime Cockrum engaged in a variety of ventures, including flat-boating, merchandising, mill operations, and railroad promotion (*History of Gibson County* 1884). Nelson, in turn, ran a farm and a nursery, contributed to the *Fort Wayne Sentinel* newspaper, worked for the federal land office, and served as a director on canal and steamboat companies (Dunn, 1914).

Others, like William Miller, spent much of their lifetime working (or managing) farms, and, on entering their retirement, devoted themselves to promoting economic development in their communities (*History of St. Joseph County*, 1880). Some agricultural improvers, like William Franklin and Samuel Herriott, were not farmers at all, but career lawyers and politicians, merchants, bankers, and hotel keepers (*Eminent and Self-made Men*, 1880; *Biographical Directory*, 1980). A few, like the abolitionist Quaker Moses Bradford, were zealous champions of reforms that were considered radical (Whitson, 1914). Most advocates of agricultural improvement, however, engaged in reform efforts that were considered to be respectable and relatively uncontroversial. Nearly all had spent the better part of their lives in Indiana's fledgling settlements. What they believed—even the most firmly committed farmers among them—was that Indiana's growth, the social improvement of her people, and the prosperity of her farmers were intimately connected. For Hoosier farmers to thrive, more people, better infrastructure, and more industries were needed. For economic development to occur, Hoosier farming had to be improved. Only on a foundation of sound economic development could social improvement spread. For any of this to occur, Indiana's agriculture had to be improved, and the state's economy could not continue to be overwhelmingly agricultural in nature.

The pioneer days were over. Samuel Perkins spared his listeners of the Marion County Agricultural Society few details of what would take their place.

> The days of early simplicity, of backwoods enjoyment and suffering, of log cabins, . . . coon skin caps, cow-path roads, exclusive horse-back and wagon traveling facilities, destitution of a market for surplus produce, and isolated living, with bare-footed girls in clearing and corn and potato patches, are passing. . . . in their place are come handsome residences, surrounded by neatly fenced yards . . . and having richly furnished parlors, china and silver ware, clothing of silk and broadcloth, beaver and silk hats, turnpike and plank, and railroads, handsome carriages, a large and growing city, furnishing a ready and steady cash market, neatly dressed children, with arms laden with books, wending their way to school, or perchance, if young ladies, singing to the music of the piano in the parlor. (Perkins, 1854–1855, p. 390)

The Agricultural Civic Mission in 1850s Indiana

Throughout human history, improvements in agriculture supplied the ingredients needed for civilizations to flourish. In a single generation, Indiana's "unbroken wilderness" had been supplanted by "rich and delightful farms" with "the produce wagon and the canal boat hourly pass[ing] by our doors." Soon railroad lines would be operating, with the panting of the "*iron horse*" and "the shrill scream of its whistle" reverberating through the countryside (Wright, 1852b, p. 81). The next phase of civilization's progress would "begin on every man's farm." Every farmer would have to be "aroused to a sense of his duty and his obligation" to take "hold of the work with a firm and invincible determination to press forward" (Durham, 1853, p. 207).

To fulfill their duty to civilization, what farmers most needed to do was to have "more self respect" for themselves as farmers (Miliken, 1852, p. 98). The conventional wisdom—endorsed by many farmers—that "a man may be fit for a farmer when he is fit for nothing else," could not remain in place. The "farming interests of the nation" were simply too important to the public well-being (Bringhurst, 1852, p. 85). It was "an axiom of political economy that the wealth of a country or community, is based upon her agricultural products." Was it not self-evident that, if commerce carried the products of the farm and manufactures refined them, that "the cultivators of the soil are the main source of wealth" (Levering, 1852, p. 344)? If so, the farming community bore the greatest burden for improving the public well-being (Baird, 1852). Farmers were engaged in the employment most essential to prosperity, and they were the most numerous portion of the population. If they failed to improve, Indiana could not improve. Farmers' agricultural practices, therefore, were "too important to be left alone, without being infused with science and knowledge" (Smith, 1852, p. 170).

In a state where approximately 20 percent of the adult population was classified as illiterate by the United States Census takers—and the preponderance of illiterates were farmers—Indiana's agricultural improvers hoped to shake off indifference to book-learning and scientific agriculture (Barnhart & Carmony, 1954). Echoing common school reformers, they hinged their case on what modern social scientists call human capital: "It is not the most fertile soil or genial clime... [and] ... is not the mines of silver and gold that can make a nation rich; it is the amount of intelligence possessed by her citizens" (Hathaway, 1853, p. 158). As did agricultural improvers everywhere, they appealed to virtually every motive force they could think of to inspire farmers to engage in self-education, to raise different kinds of livestock, and to supply a variety of crops to nearby towns.

Indiana's improvers had concerns about what Hoosier farmers were doing however, that, while not unique to Indiana, seldom appeared in the farm papers of the Northeast. Arrayed against them was a formidable antagonist: the hog. With western cities like Chicago, St. Louis, and Cincinnati booming, and access to world markets opening rapidly, pork-packing had emerged as a major industry. Fattened on corn and herded to river cities, hogs fetched $8–12 apiece. Already inclined to raise corn and hogs, Hoosier farmers grew more. By 1850, with more than two million hogs roaming farm and forest—and trotting in droves through towns every winter—Indiana had 2.3 hogs for every man, woman, and child.

Aside from selling-out and moving-on, Hoosier farmers had but "one idea" about their farms' moneymaking potential: "HOG." That was the wrong idea. It enticed farmers away "from their legitimate pursuits." Too "speedily converted into money," cash-cropped hogs plunged farmers into "an uncertain and precarious business," one that ruined three times as many men as it made. Farmers had as little business being in the hog-trade "as with horse racing or the roulette." What would serve their best interests was "an intelligent system of farming," one with ample cattle, sheep, and other livestock, as well as several types of cash crops to rotate among their fields. Only a rational system of "judicious husbandry" could make "a community of land holders truly prosperous, independent and happy" (Gookins, 1852, pp. 304–305).

Hoosier farmers, Governor Wright told them, were hurting themselves with hog-dependence. At one time, "when the hog was the only article that would command a good price in the markets," farmers had "some excuse" for relying exclusively on the corn-and-hog combination. That time ended "some years ago." With Indiana's town-dwelling population growing rapidly and outlets to distant markets multiplying, the excuse was valid no longer. Yet, hogs continued to represent one-half of the value of Indiana's livestock, earning Indiana the dubious honor of being the second-leading hog-producing state (behind Kentucky). Farmers were poorer for their hog-dependence. Exporting hog carcasses, while spending hog-dollars to purchase goods made elsewhere, their communities were the poorer for it, too (Wright, 1853, p. 217).

Governor Wright's solution was straightforward. Hoosier farmers needed to raise fewer hogs and more of other things; the people of Indiana needed to "learn to *make more at home*, and to *buy less abroad*" (Wright, 1852a, p. 217). Rather than watch the elements of prosperity flow eastward, it would be "far better to have the market transported to us and firmly established in our midst" (Conduit, 1852, p. 200). Improved, diversified, and enlarged, agricultural production would provide the raw materials for the manufacturing and processing industries that would attract more people to Indiana's towns. Improved roads would connect the towns to the countryside and to each other; new railroad lines would connect Indiana's communities to the rest of the nation. Producing nearly everything needed, Indiana's communities would be largely self-sufficient, and able to export surplus products (agricultural, extractive, and finished) to the outside world. Joined by commerce, Indiana's farmers and town-dwellers would be mutually dependent, supporting each other's prosperity. To accomplish this, Hoosier farmers had to learn agricultural improvement's greatest lesson: the necessity of bringing "the loom and the anvil into proximity with the plow" (Barnes, 1852, p. 262).

To Indiana's leading men, it was a convincing case, particularly since it did not require much public money. Urged on by Governor Wright, on 10 February 1851, four days after a new state constitution was ratified, the General Assembly passed legislation to promote agricultural improvement. Voluntary associations for agricultural education were to be established in each county; a hub-and-spoke network would be created with the formation of a central body, the State Board of Agriculture. Through its annual meetings, regulations, and, its policy-setting example, the State Board of Agriculture would focus and coordinate local efforts.

The act for encouraging agriculture laid out a plan that mixed appeals to public-spiritedness with self interest, and moral suasion with pragmatic considerations. Wealthier farmers were already improving; they needed no additional incentives. As members of agricultural societies, however, their example, demonstration, and instruction could help their neighbors improve. To encourage leading men to organize agricultural societies, a share of public funds was made available. Once at least thirty men formed a county agricultural society and raised at least fifty dollars through voluntary subscriptions, the society would be entitled to matching funds derived from the license fees charged to traveling shows and theatrical performances. In exchange, the agricultural society was to offer inducements to improving farmers to exhibit publicly their accomplishments in agricultural improvement. Honoring these farmers with premium awards—and motivating other farmers to imitate and surpass them—was the purpose of the annual agricultural fair ("An Act for the Encouragement," 1851).

The public nature of the premium award competition at the agricultural fair would encourage competitors to innovate, while at the same time providing object lessons to fairgoers. The farmer might be introduced to new crops imported from around the globe, or to improved breeds of cattle imported from the eastern states and Europe. The all-purpose Morgan horse and the fine-wooled Merino sheep might be displayed. He might see, for the first time, McCormick's reaper, Gatling's wheat drill, or John Deere's steel plow, or a host of rivals now forgotten. He could hear about the merits of different styles of drainage, planting, cultivating, and harvesting. Given the chance to learn through observation, hands-on experience, and informal conversation even the poorest farmer could learn what wealthier farmers had learned already. Farming based on book knowledge paid dividends. Using the fair's exhibits to tap into "the very love for the *practical* [so] characteristic of our people," agricultural reformers sought to bridge the gap in attitude and learning between Indiana's corn-and-hog-raising farmers and its progressive farmers and townsmen (Kinley, 1852, p. 362).

Inspired by the agricultural fair's demonstration of what could be accomplished, agricultural improvers hoped that farmers would join the county agricultural societies. To compete for the fair's premium awards farmers had to become members; once enrolled, they could take part in the monthly agricultural experience meetings. As they began remaking themselves into model farmers, they would acquire a new vision of rural citizenship, a replacement for their corn-and-hog-based rural separatism. The new vision would bring Hoosier farmers into more continuous contact with the town-dwelling life, but its diversified commercial agriculture would shelter them from the ups and downs of a single crop market. It would enlarge their profits and secure for their families all of the material comforts enjoyed by townsmen in other trades and occupations. By teaching Hoosier farmers that farming, properly understood, encompassed far more than corn and hogs, agricultural societies would teach them how to benefit from the advance of civilization. In doing so, they would anchor Indiana's prosperity on a solid foundation.

Reflections on Civic Learning

Residing in a frontier state, Indiana's leading men had no trouble identifying what was wrong with their state. Indiana lacked what the Northeast possessed: a transportation system that granted easy access to the outside world, substantial concentrations of town-dwellers and industries, and improvement-minded farmers. Their objective, practically speaking, was not to produce a replica of, say, New York, in the Old Northwest. Indiana's geography imposed hard limits on men's visions of the possible, as did its settlement history. Farmers from the upland South did not share northeastern migrants' conception of a proper community. Accustomed to backwoods settlements of dispersed farms with the occasional country store, and wary of becoming too involved in commercial dealings, most former upland Southerners warmed slowly to the notion of planting centers of commerce in their midst.

Creating a transportation infrastructure was the first civic mission related to agricultural improvement in Indiana. Through their turnpike and canal building in the first three decades of the nineteenth century, Pennsylvania, New York, and Ohio showed Indiana how to promote development. Favorable popular sentiment and capital resources were preconditions for government sponsorship of internal improvements. In the 1820s, Indiana had very little of these things. Appeals from the best men of Indiana's counties for (poor) farmers to purchase stock subscriptions or to vote for higher taxes, fell on deaf ears. For their part, men of substance and ambition from Indiana's counties frustrated each other's efforts, scattering and squandering the limited public revenue available. Their civic vision was not much broader than the farmers' civic vision. Farm families prioritized their families' needs and cared little about the broader community (town and county). Leading men prioritized their home county in matters related to development; their community of interest did not include other counties or the state as a whole.

Viewed in that light, passage of the Mammoth improvements bill in 1836 was a momentary triumph of state vision over parochialism. It was also typical of the frontier boom mentality that is usually remembered in connection with the Far West of the second half of the nineteenth century. Indiana was booming in the early 1830s. Economic growth begets growth and confidence in the future. Attracting northeastern migrants (who were better farmers, with more capital and zeal for improvement than southerners) and gaining access to eastern markets were the prizes sought by Indiana's leading men. These prizes were worth the risk of a large public debt. The risk, after all, would be borne by all of the state's residents, not particular communities, and the economic benefits would be dispersed almost as broadly. Undoubtedly, the anticipated local benefits of canals, railroads, and turnpikes inspired enthusiasm more than a statewide vision. Nevertheless, the Mammoth projects were launched by the state as a whole and firmly tied to it through contractual obligations. The legislation conceptualized a transportation system that would cover the state and be built on behalf of the people, for the people's benefit.

Just how deeply into the social structure support for the Mammoth improvements legislation permeated is unclear. Enactments of public policy were not a direct expression of popular sentiment and determining the level of farmers' support is troublesome. Most farmers did not buy county newspapers. Few wrote letters to the editors, and, had they submitted any, in all likelihood, serious objections to the improvements plan would have gone unpublished. As elected representatives, farmers took part in the debates of the General Assembly, but the debates were not transcribed. Almost by definition, these men were exceptional farmers, anyway, men who were already inclined to favor economic development and agricultural improvement. Hog-selling farmers had less reason to want canals and railroads since their principal use of such facilities occurred once a year. However, with an eye to future export sales of crops, they could have been convinced to support the proposal. After all, the infrastructure projects were supposed to pay their own way, and few farmers would bypass accessible routes on their occasional trips to market. Lukewarm support, at best, for the Mammoth improvements legislation seems to be the most reasonable verdict about the general opinion among farmers.

In the aftermath of the Mammoth projects' failure, farmers did a great deal of complaining about townsmen's excessive ambitions and corruption and taxes. Hindsight's perceptiveness flavored their grumbling, no doubt.

Be that as it may, farmers made sure the Mammoth experience would not be repeated. Delegates elected to the 1850–1851 constitutional convention went with a mission to discipline future sessions of the General Assembly. They closed the state government's purse to development schemes. At most, the public's money could be used as seed money to support improvement efforts carried out by voluntary associations and local communities. On the heels of the constitutional convention, the common school system—funded for the first time by a state property tax—was one beneficiary; the tandem of agricultural societies and fairs was another. Both were ways for the state government to contribute to the people's improvement without being directly involved in economic development.

Historians of Indiana have long pointed to the failure of the Mammoth improvements plan as a defining moment for Indiana's character: an ethos of limited government and low taxes as well as an aversion to grand schemes solidified into articles of faith. For the next two generations, in matters related to economic development, men who wanted any kind of state government action made relatively modest proposals; the legislative process toned down the proposals further yet. Beyond development, outside of funding a handful of benevolent, educational, and penal institutions, the state government took upon itself few responsibilities. It set guidelines and rules for county and incorporated town governments (notably, these were distinct entities), but left implementation in the hands of local officials. Generally, Hoosiers lacked confidence in government and looked instead to private efforts and voluntary associations. Their state legislators crafted permissive laws that encouraged that tendency by granting individuals and local groups wide latitude to pursue their goals as they saw fit.

The approach had the merit of being a flexible means of getting things done. It was well suited to minimizing public expenses and taxes, to avoiding statewide controversy, and to permitting local communities to determine their own character and fate. Its unintended consequence was to inhibit the formation of a statewide civic consciousness among Indiana's people. State-sponsored institutions and endeavors, to be sure, are not the only means of fostering robust attachments to a body politic, but they help. If nothing else, public institutions provide focal points for citizens of different localities, experiences, and outlooks to dispute what is good for the community as a whole and what is not, what everyone should or should not do. The doctrines of voluntarism and localism, on the whole, circumvent the issue of how such a broadened sense of community can be created and sustained. Indiana's people already viewed the world from a strictly local perspective. Dismantling the state government's functions—and refusing to permit it to take on new ones—crippled one potent source for building bonds of solidarity among the citizenry.

At the same time, the state government's sharply circumscribed role in society created the necessity and space for voluntary associations to flourish. Voluntary associations can foster bonds of attachment that perform functions similar to a sense of civic community. Through their organizational structures, associations can create coalitions of the like-minded, who, together, may pursue socially desirable ends. Through their activities they can teach people—explicitly and implicitly—what their place in society is and how it should govern their actions. In the 1850s, Indiana's agricultural improvement societies were intended, in part, to do these things. At the state level, through representation on the State Board of Agriculture, a broad coalition of leading townsmen and farmers took shape. At the county level, agricultural societies created similar coalitions. The meetings and fairs were intended to expand farmers' horizons, to bring them in contact with men and ideas from other places and to make them participants in a common venture. Farmers, generally, were not association-joiners and they viewed improvement of all kinds with a measure of skepticism. Advocates of agricultural societies hoped that farmers' self-interest in profitable farms would change their behavior and their minds. If it did, it would be a significant step toward improved farms, economic development, and the creation of a civic community.

References

Abbott, C. (1978). Indianapolis in the 1850s: Popular economic thought and urban growth. *Indiana Magazine of History, 74*(4), 293–315.

Act for the encouragement of agriculture. (1851). *Indiana Laws* (35th session), 6–8.

Baird, W. M. (1852). Address to the Franklin County agricultural society. In *Annual Report of the Indiana State Board of Agriculture*. Indianapolis, IN: State Board of Agriculture.

Barnes, W. A. (1852). Address to the Porter County fair. In *Annual Report of the Indiana State Board of Agriculture*. Indianapolis, IN: State Board of Agriculture.

Barnhart, J. D. & Carmony, D. F. (1954). *Indiana: From frontier to industrial commonwealth* (Vol. 1). New York: Lewis Historical Publishing Company.

Biographical directory of the Indiana General Assembly (Vol. 1). (1980). Indianapolis: Indiana Historical Bureau.

Bringhurst, T. H. (1852). Address to the Cass County fair. In *Annual Report of the Indiana State Board of Agriculture*. Indianapolis, IN: State Board of Agriculture.

Carmony, D. F. (1998). *Indiana, 1815–1860: The pioneer era.* Indianapolis: Indiana Historical Bureau and Indiana Historical Society.

Carter, H. L. (1946). Rural Indiana in transition, 1850–1860. *Agricultural History, 20*(2), 107–121.

Cayton, A. (1996). *Frontier Indiana.* Bloomington: Indiana University Press.

Conduit, A. B. (1852). Address to the Morgan County agricultural society. In *Annual Report of the Indiana State Board of Agriculture.* Indianapolis, IN: State Board of Agriculture.

Daniels, E. D. (1904). *Twentieth century history of LaPorte County.* New York: Lewis Publishing Company.

Dennis, W. T. (1850). Farming in Indiana. *The* (Albany) *Cultivator, 7*(11), 369.

Dunn, J. P. (1914). *Memorial and genealogical record of representative citizens of Indiana.* Indianapolis, IN: B. F. Bowen and Company.

Durham, M. S. (1853). Address to the Vigo County agricultural society. In *Annual Report of the Indiana State Board of Agriculture.* Indianapolis, IN: State Board of Agriculture.

Eminent and self-made men. (1880). Cincinnati, Ohio: Western Biographical Publishing Company.

Esarey, L. (1917). The pioneer aristocracy. *Indiana Magazine of History, 13*(3), 270–287.

Gookins, S. B. (1852). Address to the Vigo County fair. In *Annual Report of the Indiana State Board of Agriculture.* Indianapolis, IN: State Board of Agriculture.

Hathaway, G. (1853). Address to the LaPorte County agricultural society. In *Annual Repot of the Indiana State Board of Agriculture.* Indianapolis, IN: State Board of Agriculture.

Helm, T. B. (1881). *History of Delaware County Indiana.* Chicago: Kingman Brothers.

History of Gibson County Indiana. (1884). Chicago: James T. Tarff and Co.

History of St. Joseph County Indiana. (1880). Chicago: Charles C. Chapman & Company.

Kinley, I. (1852). Education of the farmer. In *Annual Report of the Indiana State Board of Agriculture.* Indianapolis, IN: State Board of Agriculture.

Levering, J. (1852). Improvements in agriculture. In *Annual Report of the Indiana State Board of Agriculture.* Indianapolis, IN: State Board of Agriculture.

Madison, J. H. (1986). *The Indiana way: A state history.* Bloomington: Indiana University Press.

Miliken, J. P. (1852). Address to the Dearborn County fair. In *Annual Report of the Indiana State Board of Agriculture.* Indianapolis, IN: State Board of Agriculture.

Nation, R. F. (2005). *At home in the Hoosier hills: Agriculture, politics, and religion in southern Indiana, 1810–1870.* Bloomington: Indiana University Press.

Perkins, P. (1854–1855). Address to the Marion County agricultural society. In *Annual Report of the Indiana State Board of Agriculture.* Indianapolis, IN: State Board of Agriculture.

Philbrick, F. (1965). *The rise of the west, 1754–1830*. New York: Harper and Row.

A Portrait and Biographical Record of Delaware County, Indiana. (1894). Logansport, IN: A. W. Bowen and Company.

Power, R. L. (1935). Wet lands and the Hoosier stereotype. *The Mississippi Valley Historical Review, 22*(1), 33–48.

Power, R. L. (1953). *Planting corn belt culture: The impress of the upland Southerner and Yankee in the Old Northwest*. Indianapolis: Indiana Historical Society.

Rose, G. S. (1985). Hoosier origins: The nativity of Indiana's United States-born population in 1850. *Indiana Magazine of History, 81*(3), 201–232.

Shoemaker, R. S. (1981). James D. Williams: Indiana's farmer governor. In R. G. Barrows & S. S. McCord (Eds.), *Their infinite variety: Essays on Indiana politicians* (pp. 195–222). Indianapolis: Indiana Historical Bureau.

Smith, H. M. (1852). Address to the Knox County fair. In *Annual Report of the Indiana State Board of Agriculture*. Indianapolis, IN: State Board of Agriculture.

Thornbrough, E. L. (1965). *Indiana in the Civil War era*. Indianapolis: Indiana Historical Bureau and Indiana History Society.

Turner, F. J. (1950). *The United States, 1830–1850: The nation and its sections*. New York: P. Smith.

Weik, J. W. (1910). *Weik's history of Putnam County, Indiana*. Indianapolis, IN: B. F. Bowen and Company.

Whitson, R. C. (1914). *Centennial history of Grant County, Indiana*. New York: Lewis Publishing Company.

Wright, E. W. (1852a). Address to the Carroll County fair. In *Annual Report of the Indiana State Board of Agriculture*. Indianapolis, IN: State Board of Agriculture.

Wright, J. A. (1852b). President's report. In *Annual Report of the Indiana State Board of Agriculture*. Indianapolis, IN: State Board of Agriculture.

Wright, J. A. (1853). Address to the Washington and Orange district fair. In *Annual Report of the Indiana State Board of Agriculture*. Indianapolis, IN: State Board of Agriculture.

3

Fair Frustrations

Agricultural Education as Civic Learning in the 1850s

Inviting Farmers to Join Agricultural Associations

With their vision in place, Indiana's agricultural improvers launched their campaign to uplift the Hoosier farmer. Notices were posted in local newspapers inviting farmers to assemble at county courthouses and churches. There the agricultural improvement legislation and a circular from Governor Wright were read. Prominent men entertained "with appropriate, interesting, and animated addresses" on the purposes and good results of agricultural societies (Agricultural Meeting, 1852, p. 210). Committees were formed to prepare constitutions and bylaws, to develop schedules of discussion topics, and to invite future speakers. Before departing, the members elected officers and resolved to generate interest among their townships' farmers (Agricultural Spirit, 1851).

Within two years (1851–1852) agricultural societies were organized in 45 counties, and fairs were hosted in at least 20 counties, including a State Fair at Indianapolis attended by more than 30,000 visitors. Governor Joseph A. Wright, performing double-duty as President of the State Board of Agriculture, sounded a triumphant note: the political campaign season

Civic Learning through Agricultural Improvement, pages 45–71
Copyright © 2011 by Information Age Publishing
45

notwithstanding, "the prevailing subjects of interest among the multitude were connected with the advancement of the cause of Agriculture, and the encouragement of the various branches of useful labor." The work accomplished to date had "satisfied the public mind, beyond doubt, that these associations are well calculated to promote the cause of agriculture, mechanics, and manufactures" (Wright, 1852, pp. 1, 6).

In counties throughout the state, advocates of agricultural improvement were less convinced of the people's receptiveness to agricultural societies. In some counties, organizers had difficulty meeting the law's requirement to enroll at least 30 members; only "untiring zeal" enabled "a few energetic men" to establish their agricultural associations ("Hendricks County Report," 1853). Starting its second year, the Putnam County Agricultural Society was forced to repeat their initial organizing strategy. They appointed one man to each township with instructions to enlist "four active friends of agriculture" to organize a "mass agricultural meeting" ("Putnam County Report," 1852, pp. 217–218). In Putnam County, as elsewhere, many "very good and thrifty husbandmen" continued to withhold their "influential support, or material aid" from agricultural societies (Barnes, 1852, p. 264). If the most public-spirited men in 45 of Indiana's 92 counties were awakening to the cause of promoting agricultural improvement among the farmers, a good deal of prodding was involved.

Although reluctant to join associations, prominent farmers were inclined to support the intentions of agricultural improvement, at least. Among ordinary farmers, hostility and apathy prevailed. In southern Indiana, agricultural societies met "a strong tide of opposition" from "superstitious enemies" who were "opposed to *book-farming*" ("Ohio and Switzerland County District Report," 1853, p. 168). In the west, the farmers' "time-honored prejudices" made for "considerable labor and vexation" ("Tippecanoe County Report," 1851, p. 189). After only two years, one organizer in eastern Indiana conceded defeat. The scrub cattle and "scrub farmers" of his county would "only be removed by emigration or death" ("Fayette County Report," 1853, p. 85). Northern counties fared little better. Their agricultural societies were patronized by men "mostly engaged in other pursuits" ("Elkhart County Report," 1851, p. 57).

Why were not farmers interested in agricultural societies? Rural traditionalism and mistrust of innovation played its part; this was the preferred explanation among agricultural improvers. As an agenda, though, the message of agricultural improvement carried more than farm profits, and its messengers tended not to notice how they interfered with the message. Unlike common farmers, most organizers were well educated and well off; many were town businessmen (active and retired) who owned

farms; and, many hailed from northeastern states. Coming from new arriv-
als, town-dwellers, and wealthy men whose hands stayed clean when they
engaged in farm work, the agricultural improvement message smacked
of regional chauvinism and class condescension. Forthright talk of scrub
stock and slipshod farming, undoubtedly, pushed away many common
farmers (Blanke, 2000).

The educational style of the agricultural meetings did little to inspire
enthusiasm. Wealthy gentlemen who dabbled in fancy farming were often
accused of not taking seriously the business side of farming ("Fancy Farmers," 1852). Agricultural meetings were too serious, taken up with lectures,
the presentation of communications, and committee reports. If men knowl-
edgeable of the business side of farming predominated, discussions were
overwhelming practical. A typical meeting discussed at length the comparative advantages of driving cattle to market or shipping them by rail; the design and manufacture of plows; and, methods for cultivating and processing
flax ("Tippecanoe County Report," 1851). If the assembly was filled with
men who lacked practical farming experience, the advantages of agricultural chemistry and educating boys in farming were discussed ("Franklin
County Report," 1853). The first type of meeting, at least, offered some tips
for working farmers. Envisioned as a "plain, practical *conversational* series of
meetings," the societies offered a fare of improvement that was beyond the
comprehension—or the inclination—of most farmers ("Decatur County
Report," 1852, p. 103). Even a contributor to the recently launched agricultural newspaper, the *Indiana Farmer*—almost, by definition, a supporter
of agricultural improvement—thought that the meetings promised "to be
rather a dry business" (Weir, 1852, p. 152).

Encouraging Farmers with Fair Premiums

Since they were unable to entice farmers to attend agricultural meetings,
members of the agricultural societies approached their first fairs with mis-
givings. Skepticism, caution, and desperation coexisted. Directors were
"earnestly advised" by friends to "abandon the enterprise." Was it not "silly
and hazardous" to incur the "heavy liabilities" of hosting a fair when so few
farmers seemed interested ("Decatur County Report, "1852, p. 103)? The
agricultural societies had no alternative but to host fairs. Their "prospects
for usefulness, if not for life itself," hinged on their first fairs' ability to attract farmers into the associations ("Elkhart County Report," 1851, p. 57).

The earliest fairs in Indiana were modest events. Some were hosted in
the county seat's public square. Tables covered with vegetables, handicrafts,
homemade cloth goods, and a jumbled assortment of smaller articles filled

the courthouse. The square itself was fenced-in to prevent livestock from browsing in town-dwellers' gardens and filled with pens for cattle, hogs, sheep, and horses, a show ring, and narrow lanes for foot traffic. Other fairs were hosted on farmland lent for the occasion. The Greene County Agricultural Fair was situated half in the woods and half in a potato patch. The grounds had "no accommodations at all" and its program had no system: any person who brought livestock or articles "was allowed the privilege of poking them into any corner he could find" (*History of Greene and Sullivan Counties*, 1884, pp. 62–63). Other fairs were better organized, but most displayed abundant evidence that the agricultural fair was, indeed, an innovation in Indiana.

Given Indiana's frontier conditions and the novelty of agricultural fairs, the early fairs' displays were somewhat wanting. Only one long-wooled and one fine-wooled sheep were exhibited at Monroe County's agricultural fair ("Monroe County Report," 1857). In Elkhart County, "not a single pure bred animal" could be found; the prize bull was "a quarter blood Durham, unhitched from a breaking plow" for the occasion (Bartholomew, 1930, 116). Embarrassed by this, the Board of County Commissioners charged a member with the task of importing a purebred Shorthorn bull prior to the second fair. Importing purebred livestock into frontier communities was costly, therefore, specimens of the fancy breeds were hard to find. Farmers who were willing to display crops at fairs were almost as rare. Had it not been for "contributions from the city [Fort Wayne], there would scarcely have been any show" of fruit, vegetables, and grains at Allen County's fair. Only one farmer from the country brought fruit to display, a "rule that will hold true in regard to everything else" ("Allen County Agricultural Society," 1855).

Despite the meager displays, the early fair inspired great enthusiasm. "If any were in doubt" about the fair's impact, "their doubts would speedily be removed if they could hear the general inquiry prevailing in all corners of the county after the best breeds of stock and the best labor-saving machinery" ("Elkhart County Report," 1851–1852, p. 108). One observer was confident that, since its first fair, farmers imported into his county more improved livestock and farm implements "than had ever before been owned here" ("Rush County Report," 1853, p. 190). Impressed by what they witnessed, southern Indiana's farmers were "beginning to see the folly of destroying their lands by cultivating corn and raising hogs" ("Washington and Orange District Report," 1853, p. 214). The earliest fairs did not quite deliver "the death blow" to "old fogyism" in agriculture ("Bartholomew County Report," 1853, p. 46). They did, however, mark the turning point in the lives of the agricultural societies.

Inspired by the interest fairs generated, agricultural societies moved to increase it. At the initial fairs, agricultural societies appealed primarily to personal honor and public spirit to encourage people to participate in the competitions. Only diplomas, a handful of small cash awards (typically ranging from 25 cents to one dollar), and the occasional silver goblet were distributed as rewards. With successful fairs behind them, agricultural societies could do more. First-place winners in important categories had silver pitchers, goblets, and bowls bestowed upon them in elaborate ceremonies. The "*silver* encouragement" caused "the gardener to give his vegetables an extra dressing" and made "the farmer plow deeper the soil" ("Fayette County Report," 1853, p. 84). More aptly, it encouraged wealthy men to import into Indiana the $1,500 horses, the $1,000 bulls, and the $800 jacks, the purebred varieties that would be used to improve the common stock. One round of experience with fairs that boasted awards of silver trophies was sufficient for leaders of agricultural societies to reach the obvious conclusion: If fairs failed it would be because they did not furnish sufficient incentives to competitors. When the numbers and sizes of silverware prizes increased, the "results fully justified their anticipations" ("Franklin County Report," 1854–1855, p. 35).

The typical Hoosier farmer was unlikely to beam at the prospect of being handed a silver bowl in a public ceremony. The trophies were not intended for him. They were for the wealthiest men. Distinctive for their character and accomplishments, prominent men were supposed to show lesser and younger men to what sorts of things they should aspire. Lesser men were supposed to emulate, that is, to imitate and strive to match, their betters to the extent permitted by their talent and treasure. No one expected ordinary farmers to compete against gentlemen farmers and wealthy farm owners. What dirt farmer would place his long-legged, long-snouted hog that roamed the woods alongside a finely proportioned Berkshire hog that resided in a comfortable barn? Who, in the midst of stumps and weeds, thought he could win a prize for the best cultivated farm? Common farmers stood only to be humiliated—not honored—by competing for first-place prize premiums.

Accustomed to the distinction between the best men and those who emulated them, agricultural societies created a two-tiered award system. Instead of silver trophies, second- or third-placing farmers received awards to help them "study their profession." Some men received "the best agricultural and horticultural works both east and west" ("Lagrange County Report," 1856, p. 219). Others were awarded the State Board of Agriculture's *Annual Report*, which was distributed free to agricultural societies, or a one-dollar subscription to a year's worth of the *Indiana Farmer*. Books and farm papers were not the rewards best calculated to appeal to working farmers. When given the choice, most "of those whose award was the *Indiana Farmer* took one dollar instead" ("Marshall County Report," 1859, p. 74). By 1857 only ten agricultural societies reported awarding any books or farm papers as premiums (State Board of Agriculture, 1857a, pp. 713–714).

Despite the reading incentive's rapid decline, the fair-going experience may have inspired farmers to study their profession. Judged by the agricultural press's explosive growth in the 1850s, many farmers did begin reading about agricultural improvement (Gates, 1960). Far fewer were inspired to participate in agricultural society meetings. Agricultural societies liked to boast, but their membership figures were an illusion of popular support. Despite having 170 members, for example, the Marshall County society's regular work was "generally performed by ten or twelve persons" ("Marshall County Report," 1859, p. 77). Likewise, Warren County had 300 members, but few showed "little interest" beyond paying the one-dollar annual membership fee ("Warren County Report," 1854–1855, p. 164). The fee entitled a man to join an agricultural society; it also gained his family admission into the fair. The fair was supposed to be the entry point into the agricultural societies, but most farmers failed to follow up on it.

Indeed, only a small proportion of farmers took part in the fairs' competitions. Agricultural societies' complained constantly about the poor displays of "*agriculture proper*," that is, in the grains, vegetables, and smaller livestock raised by general farmers ("Hendricks County Report," 1853, p. 116). Awarding fewer books and more one-dollar prizes did increase farmers' participation, though, by the late 1850s, as indicated by a new complaint that surfaced in letters submitted to the agricultural press. Was it not obvious that wealth gained from non-farming sources gave gentlemen farmers an insurmountable advantage that discouraged other farmers from competing? The same point was made indirectly, by denouncing the "favoritism" shown toward "a few wealthy exhibitors" by competition judges ("Bartholomew County Report," 1857, p. 120). Letter-writers also demanded an "equalizing" of the premium awards. Their grains, vegetables, and implements were the results of labor and intelligence, not simply accumulated wealth, and were worthy of same honor given to fancy cattle (Innis, 1859, p. 6). The target of the complaining was the wealthy gentleman farmer. Working farmers, men who were distinguishing themselves from their peers through self-improvement, were demanding a share in the public honors awarded at the agricultural fairs.

Faced with the complaints, the State Board of Agriculture distributed greater sums of money across neglected categories of the State Fair's premium list (State Board of Agriculture, 1858, p. xviii). Typically, the State Fair's premium list served as the model followed by county societies, but in this case the action was taken too late. Leading men across Indiana's counties had reached a different conclusion about how best to reform the competitions. Originally intended to stimulate friendly rivalry among local farmers, the early fairs were closed to competitors from other counties. Each county, however, had only a handful of wealthy farmers who could import fancy livestock. After the initial fairs, they found little honor in winning closed competitions. Instead of adopting reforms to increase the participation of working farmers from local townships, agricultural societies declared their fairs open to the world (Fayette County Report, 1854–1855). This was an invitation for the wealthiest individuals of Indiana's counties—with interests in all conceivable pursuits—to attend each other's fairs. The move may have been a legitimate way to increase the competitiveness of the fancy cattle show, but by bringing the wealth advantage to bear upon other premium categories, it pulled the competitions out of the reach of most working farmers.

Throwing open the fairs signaled the end of efforts by most prominent farmers to bridge differences between themselves and run-of-the-mill farmers. Members of the agricultural societies had a ready justification for the change, however. The people competing were not the intended learners. For competitors, the annual event was "examination day," the occasion "where the lessons of life, as developed by another year's study and experience are recited." Through the exhibits and displays—and the judging and discussions surrounding them—the agricultural fair achieved its most potent influence upon its real intended learner, the body of people who came for the show. With the "exhibitor as a teacher," the thousands who gathered could learn more than any "cartload of reading" could supply; well informed and well inspired, they could return home to improve in their calling (State Board of Agriculture, 1859, pp. x–xi). If the fair's educational potential lay in its ability to bring local farmers into contact with the exhibits, only the biggest fairs, with the largest premium lists, and the very best livestock, machinery, and manufactures would do.

Civic Learning from Fair Hosting: The Town "Takeover" of the Agricultural Society

The agricultural fair, not the society, had become the object of affection among prominent farmers. His county's leading men, one agricultural society secretary complained, had become "lukewarm in their attachments, if not totally indifferent" to the agricultural society's educational work. Instead, they seemed "to measure the benefits of the Society from the amount of premiums drawn at the fair" ("Knox County Report," 1854–1855, p. 77). Prominent farmers elsewhere displayed "no other interest in the welfare of the Society than to attend the annual exhibition, bringing with them some article which owes its superiority to the wealth and generosity of Dame Nature" ("Marshall County Report," 1859, p. 73). Trophies and public ceremonies bestowed sufficient honor to persuade leading men to display their purebred cattle, horses, and other livestock. For gentlemen farmers who dabbled in the agricultural arts on country estates, the fair's competitions were a form of conspicuous display, a way to show off their wealth. For ambitious farmers who were seeking to grow their fortunes, the public competition was a way to become recognized as importers and breeders with animals to sell. To achieve these purposes, agricultural discussion meetings were not necessary.

Outside the fair, prominent farmers ceased trying to draw common farmers into the monthly meetings. In 1852, the Franklin County society could "scarcely see how a society can discharge the duty it owes to community without holding frequent meetings." Three years later, with "some doubts as to its propriety," they revised their constitution to hold meetings only four times each year ("Franklin County Report," 1852, p. 113, 1854–1855, p. 34). The Delaware County Agricultural Society did the same, after finding it "impossible to interest the same persons in two popular movements" during the election year of 1854 ("Delaware County Report," 1854–1855, p. 15). Active in local, county, and state politics, involved in multiple business interests, and occupied with their own farms, the leading men of Indiana's counties had too many demands upon their time to attend agricultural meetings or to "lend a helping hand" ("Marion County Report," 1854–1855, p. 111). They, personally, derived little educational value from experience sessions, and unless other farmers attended there was no one to instruct. The monthly meetings were abandoned. If ever an agricultural society combining features of the scientific society and the lyceum had a chance to flourish in Indiana, that day was past.

The agricultural society's sole mission was to host the agricultural fair. Public enthusiasm for the annual event took what was left of the agriculture out of the agricultural society. Hosting a single fair took a great deal of manpower and resources. Months in advance, premium lists had to be devised, advertisements written and posted, and supplies ordered. In the days leading up to the fair, the ground for show rings had to be worked, fences and pens erected, and stands built for exhibits, food-selling, and crowd-seating. During the fair, men had to serve on award committees; others were needed to oversee the exhibiting, still others to collect fees and fill out entry forms, to manage the crowds, and to police the grounds. At the fair's end, the grounds had to be cleared and, until land ownership became possible, the grounds had to be cleared and the facilities torn down.

As fairs expanded in scope and scale the difficulties grew, and the agricultural societies' capabilities became strained. Public faultfinding mounted: with inadequate space and facilities, with hastily improvised award committees and incompetent judges, with what was and was not included in the premium list, and with the allocation of prize money. The labors involved in fair hosting were "the lot of the very few, who stand the blunt of all calumny, and receive, as a general thing, but little credit for their efforts" ("Posey County Report," 1859, p. 101). Some complaints, surely, could have been avoided with closer oversight during the fair. But, to be prevented, most problems required the kind of foresight made possible only by the involvement in planning of the variety of people who were interested in the fair's competitions. Fair-hosting simply placed "entirely too great a burden of responsibility upon the officers" of the agricultural societies ("Hendricks County Report," 1858, p. 159). Yet, few people were "aware of the care and labor required . . . to render the enterprise acceptable to the great mass of the exhibitors and visitors" (State Board of Agriculture, 1857b, p. 4). Being confronted with complaints is a common-enough experience among those who organize events for the public. The men who led agricultural societies, however, were accustomed to approbation and admiration, not grumbling.

By itself, the public's failure to show proper appreciation might have been sufficient to stoke resentment among prominent farmers. It was compounded by the fact that they were personally financing the fairs. Most of the earliest fairs were hosted on privately owned farms, grounds lent, often, by officers of the agricultural society. The costs of fencing, building materials, and labor were often provided by the land owner or shared among several members of the agricultural societies. Agricultural fairs, moreover, were not paying enterprises. Revenues covered only one-third to one-half of the expenses; the shortfall was covered through voluntary contributions. Torn between wanting to expand the premium lists (and thus generate more public interest) and not wanting to be responsible for debts, agricultural societies eked out a precarious existence. At the close of 1856 only seven county societies had more than $150 in the treasury, seven were in debt, and the rest were on the margin of subsistence (State Board of Agriculture, 1857a, pp. 712–714).

The years 1855 and 1856 were dismal years for agricultural societies. Poor weather during the fair season depressed attendance and revenues;

the people's refusal to contribute to planning and operations took its toll. Finding that the "few persons" who had organized its first two fairs were no longer willing to "make the personal exertion," Miami County abandoned its fair in 1855 ("Miami County Report," 1857, p. 39). Faced with the "continued indifference of the mechanics and farmers," the Huntington County Agricultural Society "was suffered to go down" in 1856 ("Huntington County Report," 1857, p. 132). Rainy weather caused the Boone County Agricultural Society to cancel its fair in 1855; the fair was cancelled the next year when too few people agreed to plan for it ("Boone County Report," 1857, p. 158). Hosting successful agricultural fairs imposed too great a "tax" on the "time and pockets" of prominent farmers ("Rush County Report," 1856, p. 255). Rather than continue to shoulder the burdens of fair-hosting, they were willing to let the agricultural societies and fairs fold.

Despite the dismal experience of late, the president of Dearborn County's agricultural society remained hopeful that "a few appropriate remedies" might infuse new life into his agricultural association ("Dearborn County Report," 1854–1855, p. 177). The conclusion drawn in Porter County, perhaps, sums up best the situation and the remedy. The "plain farmers" of the county had "taken hold of the handles and made as straight a furrow as [they] could under the circumstances," but experience proved "that an agricultural society cannot flourish without the aid of professional men" ("Porter County Report," 1853, p. 174).

Whatever their interest in agriculture proper, leading townsmen (merchants, bankers, and businessmen of various kinds) were not willing to let the fairs go down. The financial benefits of hosting fairs were too great to forego. A successful fair doubled or tripled (or more) a town's population during fair-week by attracting farm families from the outlying regions ("Posey County Report," 1859). Farm families welcomed the opportunity to come into town, to socialize with other members of the community, to be entertained by the fair's amusements, and—while wallets were fat from the fall harvest's sale—to purchase all the manufactured goods and ready-made clothing the town merchants could supply. Just how many farm families left town with a sewing machine or other goods in their wagons, and how many agreements were struck for the delivery of so many bushels of farm produce is unknowable. It is certain, though, that the agricultural fair made good on one promise Governor Joseph Wright made to the General Assembly: it revealed the "real wants of the people, and the prospect and means of supplying those wants" (Wright, 1851b, p. 10).

In the 1850s and later, prominent farmers claimed that townsmen took over the agricultural societies by holding secret meetings to elect new officers or by taking advantage of poorly attended meetings to reorganize. This sort of thing appears to have happened in some cases. When Morgan County's agricultural society canceled its 1856 fair at Martinsville, a rival faction launched a fair at nearby Monrovia ("Morgan County Report," 1856). In March of 1858, a "special meeting" of the St. Joseph County Agricultural Society elected a new slate of officers, even though, typically, elections were held in December or January (*History of St. Joseph County*, 1880). In most cases, however, the town takeover of agricultural societies was done openly, through appeals for support in county newspapers and open canvassing among prominent citizens ("Boone County Report," 1857; "Miami County Report," 1857). Whatever the situation in a particular county, when the fairs hit hard times, fewer farmers moved to aid them than men engaged principally in other pursuits.

In all likelihood, the prominent farmers of Indiana's counties willingly turned over the leadership of agricultural societies to townsmen. Fairs failed to draw farmers into meetings and the charm of silver cups was wearing thin. Few gentlemen farmers relished the role of fair manager. The work involved was better suited to younger businessmen than to retired men in the later years of their lives. The deciding factor was straight forward. There was not a great division between town and country among the active members of agricultural societies. Whether they were prosperous farmers, retired gentlemen farmers, or representatives of town businesses and professions, all were members of the rural elite. They agreed that the agricultural fair improved the farmer and that hosting it improved the town's commercial activity. At the time, neither the priority order nor the question of who took responsibility for accomplishing these aims mattered all that much. In later years, however, the leadership transfer would have significant effects on the agricultural fair, as the aim of improving the farmer became dwarfed by the aim of boosting the town-based economy.

Using Fairs to Grow Towns in the 1850s

Growing towns had always been part of agricultural improvement's mission. Encouraging farmers to scale-up and refine their subsistence activities for commercial sale to nearby towns was intended to supply needed foodstuffs and the raw materials for fledgling industries. Since it brought farmers into town, the agricultural fair offered merchants the opportunity, typically unspoken, to profit from selling supplies and consumer goods. A successful fair could also promote economic development by publicizing a county's resources and advantages. New settlers and investment capital were needed for the local manufacturing that could replace the industries of eastern states and Europe in meeting local farmers' wants. Closer commercial collaboration between local farmers and townsmen could keep their dollars at home, where they could be reinvested to improve the infrastructure, industries, and institutions of Indiana.

When he stumped the state in the early 1850s, Governor Joseph Wright urged people to take aggressive efforts on behalf of this agenda, his "local and *Home Policy*" (Wright, 1853, p. 220). The kind of improvement he desired could be seen in Wayne County. On his first visit there, Governor Wright found, to his surprise, that the "extensive collection" of "grain, stock, carriages, wagons, threshing machines and other farming implements" rivaled those at the State Fairs of Ohio and New York. At their agricultural fair, the people of Wayne County provided a vivid demonstration of "the great benefits of placing the manufacturer and consumer side by side." That was the sound and true doctrine established by Thomas Jefferson (Wright, 1851a, p. 255). If other counties followed suit, Indiana would become "the very first State in the Union in all that makes a people happy: *Light taxes, no debts, an economical government, a prosperous, agricultural, manufacturing and mechanical State*" (p. 249).

Wayne County was somewhat unusual. Located in the Whitewater River Valley on Indiana's eastern border, it contained some of the older settlements and was densely populated with a large number of former Pennsylvanians and New Yorkers. It also led the state in manufacturing. Yet, Wayne County remained predominantly agricultural. Richmond, the county seat, was a small town; the county's residents outnumbered its 1500 inhabitants by nearly 20 to 1. Wayne County was moving rapidly toward becoming a self-sufficient community by filling-in its territory with small farms; by exporting its fair share of hogs, wheat, and timber to the world; and, by building-up its manufacturing industries. With a broad economic foundation, Wayne County supplied most of its people's wants and maintained a good balance of trade with the world. It exemplified Governor Wright's "local and *Home Policy*" (Wright, 1853, p. 220).

Most of Indiana's counties lacked Wayne County's fortunate circumstances, and few people shared Joseph Wright's sense of urgency about Indiana's quasi-colonial relationship with the East until his two terms as Governor ended in 1857. Their ambivalence stemmed, in part, from the return of agriculturally based prosperity in the late 1840s, a prosperity that was spurred by crop failures in Europe, the growth of Eastern cities, and the opening of Indiana's interior to commerce. The completion of some of the Mammoth improvements projects helped; more important, though, was the building of railroad lines throughout the state. From less than 50 miles of track in 1849, Indiana's railroads were extended to more than 2160 miles by 1860. As lines connected country villages to larger towns and cities to the world, the growing export sales of flour, meat, and lumber swelled farm profits. On the import side, a flood of (comparatively) cheap manufactured goods poured into the state. On this narrow economic foundation of long-distance exchange, Indiana's town-dwelling populations doubled: more millers, merchants, and middlemen to handle the farmers' commerce (Barnhart & Carmony, 1854).

Most of the surplus wealth generated by the enhanced economic activity followed the hogs, logs, and flour out of Indiana. Hoosiers failed to heed Governor Wright's admonition to promote aggressively local manufacturing. During the 1850s, manufacturing in Indiana did increase: it more than doubled. For gauging the trend toward economic dependency upon the East, though, the mix of manufacturing industries was more important than the size. The expansion of farmers' and townsmen's traditional economic activities (low value-added processing of agricultural and forest products) dwarfed new ventures in textiles, machinery-making, coal extraction, limestone quarrying, and other non-agricultural industries. Reliance on agricultural exports increased. The new railroad lines brought prosperity and ended frontier-enforced deprivation; in doing so, however, the railroad made Indiana more dependent on the Eastern states. The financial panic of 1857 drove home the point. Railroad lines and lines of credit tied Hoosiers' livelihood inextricably into the economic decisions made in other places (Mahoney, 1990). Faced with this reality, a growing number of Indiana's townsmen and prominent farmers became convinced that Governor Wright's economic prescription had been sound.

The change of heart was evident in the Indiana General Assembly's reversal of opinion with regard to sponsoring a Geological Survey of the state's natural resources. Intending to bring Indiana's advantages to the attention of people who might be induced to invest or relocate, the State Board of Agriculture pushed the idea at every legislative session. Judgments of indefinite postponement and inexpedient at present greeted their efforts until the financial panic of 1857 demonstrated the state's economic vulnerability. Previously, the General Assembly "avoided action by averring that no expression of public sentiment" existed in favor of a geological survey (State Board of Agriculture, 1858, pp. xii–xiv). Circulating a petition among the county agricultural societies achieved the desired effect. The next legislative session authorized a $5,000 appropriation for the geological survey with an overwhelming show of support (Act authorizing the State Board of Agriculture, 1859). For the first time since the Mammoth improvements projects, Indiana's government was opening its purse to promote economic development.

Behind the support for the Geological Survey lay a straightforward lesson: to buffer the effects of regional consolidation—and to take advantage of its opportunities—the local economies of Indiana's counties needed to be diversified. Local manufacturing industries were essential. Since railroad lines now yielded ready access to the outside world, natural resources could be exploited for export sale. If the capital and labor were available, new industries could be developed for refining the raw materials of the farm, forest, and mine into high value-added finished products. To market these goods, Indiana's townsmen need not rely only upon the demand furnished by local farmers. They could sell locally produced finished goods at home and abroad. On a broadened and more stable economic foundation, Indiana's communities could prosper. For this to occur, though, capital, labor, and initiative would have to be imported from outside Indiana.

The consolidation of the regional economy through railroad building also conveyed Governor Wright's sense of urgency to Indiana's prominent townsmen and farmers. The clinching argument in favor of aggressive promotion, though, had less to do with the economic relationship between Indiana and the East than with emerging economic rivalries within Indiana. Spurred by railroad-borne commerce, most of Indiana's towns doubled in size between 1850 and 1860. By the close of the 1850s, Indiana had 30 towns that possessed between 1,000 and 2,500 inhabitants, as well as 13 larger cities strategically located throughout the state (Barnhart & Carmony, 1954). The first to build railroad lines, the larger cities grew the fastest and the most. As their populations, manufacturing industries, and mercantile establishments grew, their influence overlapped the periphery zones of smaller cities. Each small town was competing with at least one larger city for the ingredients of economic growth: capital investment, manufacturing, and the farmers' commerce.

The so-called town takeover of the agricultural fair coincided with the crest of railroad building and the onset of economic stagnation. With the railroad's aid, the agricultural fair could serve as a short-term and long-term economic growth engine. The crowds and displays of the earliest fairs had been small; the opening of a county under the railroad's influence grew both. What the railroad brought into Indiana, the fairs put before the farmers: factory-made clothing and household items, improved farm machinery, and virtually all of the manufactured products of the commercial economy. Living in frontier communities, farm families were accustomed to material deprivation. The railroads supplied the fair and surrounded farmers with abundance and the call to spend. The single week of enhanced economic activity that a fair provided meant the difference between bankruptcy and solvency for town merchants. At the same time, with railroad lines crossing most counties, a thriving fair attracted visitors from throughout Indiana and the surrounding states. Some came only to attend the fair, but others were seeking potential home sites or investment opportunities. That was of great importance to the long term prospects of the economies of Indiana's towns.

To host the kind of fair that attracted public attention, agricultural societies needed land. From the start, members of agricultural societies had complained about the difficulties of holding fairs on borrowed ground: cramped quarters, bickering landowners, the labors involved in building (and then tearing down) fences and sheds. Soon after the General Assembly granted agricultural societies the privilege of owning up to 20 acres of land, they were rapidly converted into joint stock companies ("Act Authorizing County Agricultural Societies," 1855). Some set their fund-raising sights low—Kosckiusko County raised $395 and Hamilton $200 through the sale of $5 shares ("Kosciusko County Report," 1857; "Hamilton County Report," 1858). Others exceeded expectations. Hoping to buy 10-15 acres, the Floyd County Agricultural Society sold enough $25 shares to purchase 62 acres ("Floyd County Report," 1858). Some made long term leasing agreements with land owners, others struck bargains with the County Commissioners. Throughout Indiana in the late 1850s, as shares were sold, land selected, wells dug, and display halls erected, the aim was the same: to place the fairs "upon a more permanent basis, with a prospect of a greatly increased circle of influence and usefulness" ("Vigo County Report," 1856, p. 267).

Through common consent and indifference in the early 1850s, fairs were held at the county seat or rotated between towns on an annual basis. By the decade's close, the economic advantages of hosting fairs ended the reciprocal arrangements. Towns bid against each other, offering land, money to pay premium awards, and improved fairgrounds facilities to agricultural societies. The Switzerland and Ohio District Fair alternated between the two county seats for several years; the promise of a larger site removed it to the aptly named town of Enterprize ("Switzerland and Ohio District Report," 1859). In Warren County, the citizens of Attica "very liberally and generously donated" $1500 to the agricultural society to prevent the fair from returning to Fountain County as scheduled ("Warren and Fountain District Report," 1856, p. 295). In Dearborn County, the citizens of Aurora launched "an independent opposition fair" in 1859 hoping to attract farmers to their town instead of the county seat at Lawrenceburg ("Dearborn County Report," 1859, p. 24). Citing distance and geographical barriers as their reason, new agricultural societies were created, combining townships rather than counties. Which town would host the fair? Who would host the biggest and the best? On the eve of the Civil War, these questions were just starting to become important; their importance would grow.

With the agricultural fair assuming its new role as a town's economy booster, putting all the industries and resources of the county on display—and drawing crowds—became driving purposes. In a turnabout from earlier fairs, the secretary of the Madison County Agricultural Society was quite pleased with the cattle, sheep, horses, and hogs on exhibit. The "display of manufactured articles," however, was "rather light" ("Madison County Report," 1856, p. 277). If the farmer should display his livestock, why should not the merchant show his goods and the mechanic the tools of his trade? The mechanics of Dekalb County, in contrast, provided the "crowning feature" of their fair: a "Mechanical Tableau" composed of a 200 foot-long shed in which blacksmiths, gunsmiths, coopers, and shoemakers worked their trades in one-hour shifts ("Dekalb County Fair," 1859, p. 2). If a "fair representation from every branch of industry" was desired, a broadened premium list was needed ("County Fair," 1856, p. 2). Boots and shoes, carriages and buggies, tinware and leather goods joined the agricultural products on the list. The wider the range and the better the quality of articles on exhibit was sure to convince observers "that the right spirit exists in our county" ("LaPorte County Report," 1859, p. 62).

The desire to compete and to display the products of their toil was not restricted to the men of the towns. Was it not "a notorious fact that while a *man,* for very little extra attention to an animal, often receives a premium equal in value to the animal itself," "a *lady* is considered sufficiently remunerated for weeks and perhaps months of intense labor spent on some article of domestic manufacture, by a diploma, thimble, or at most, by some monthly periodical" ("Hendricks County Report," 1854–1855, p. 51)? To gain the town-dwelling ladies' participation, agricultural societies rewarded them more justly and erected large halls so that they could display their talents. In the Domestic or Floral Hall, cakes, "preserves, jellies, canned fruits, jams," butter, bread, and other table articles were "exhibited in the greatest abundance." Together with the "carpetings, the flannels, hosiery, linens, and other articles" on display, they "showed the taste, industry, and domestic worth of the frugal wife or daughter" ("St. Joseph County Report," 1857, p. 158). The products of the needle and floral arrangements "elicited many flattering expressions from the admiring crowds." Given proper encouragement, the town-dwelling ladies' contributions soon outnumbered, by far, those of the gentlemen and the "great mass of spectators were attracted by and pleased with" the expanded ladies' department ("Dearborn County Report," 1858, p. 55).

Although some of the articles displayed by townswomen at the fairs were intended for commercial sale, most were not. Encouraging women to participate had more to do with boosting attendance. The ladies' department "always presented something novel and exciting" ("Miami County Report," 1857, p. 143). Their talents on horseback did even more to draw in the crowds. The leading ladies of Fayette County competed for an expensive saddle ornamented with silver-enameled leather and silk fringe ("Fayette County Report," 1854). Women and girls elsewhere competed for silver goblets and spoons and gold thimbles. Most often they staidly exhibited equestrian merit but, on occasion—to the enthusiasm of onlookers and the consternation of agricultural editors—the young ladies wheeled and cut and raced each other for time. Where the former style prevailed, judges experienced great difficulty deciding the outcome. Where the more vigorous competition prevailed, the agricultural society's treasury was the sure winner ("Shelby County Report," 1854–1855; "Women Riding at the Fairs," 1855).

When the popularity of ladies' riding demonstrated what brought the crowds to the fairs, the horses moved from the show ring to the time ring. Other horse- qualities were forgotten: only speed mattered. Whether trotting, pacing, or outright racing in saddle or harness, horse speed stirred more excitement than even the rowdiest political campaign. On the first day of the 1859 Warrick County fair about 1,000 people attended; the third day "was emphatically the *day of the Fair*"—more than 5,000 people showed up to see the trotting. An unparalleled success, the fair was extended an extra day, at the request of citizens who agreed to sponsor premiums for additional races ("Warrick County Report," 1858–1859, p. 291). That same year, the introduction of horse racing achieved equally impressive results in LaPorte County. More people gathered at the speed trials than had ever before assembled at one place in the county. Lack of a high fence prevented the agricultural society from fully capitalizing: "about as many witnessed it from the outside of the enclosure as within" by standing in their wagons ("LaPorte County Report," 1858–1859, p. 177). Successful at bringing the farmer to the fair, agricultural societies set out to get him inside. Time rings one-fourth or one-third of a mile in length were built (along with higher fences), and the speed premium was made an official part of the program.

In response to the spread of horse racing, the State Board of Agriculture condemned the speed premium as "impolitic, immoral and unwise" and "against the best interests" of the State (State Board of Agriculture,

1858–1859, pp. xii–xiii). A few county agricultural societies reaffirmed their commitment to the "legitimate objects and aims" of agricultural improvement ("Washington County Report," 1857, p. 164). Most, however, were confident that "a little innocent amusement" might be "combined advantageously with the more serious business of the occasion" ("Miami County Report," 1857, p. 143). Why not use the public's enthusiasm for horse racing and other attractions to advantage? With its premium competitions, agricultural and mechanical exhibits, and addresses providing serious education, the agricultural fair was superior to "the more irrational and barbarous entertainments" found in other countries. Surely a well-managed fair could function as both "a school of instruction and a source of amusement, inspiring all with encouragement and honorable emulation in the legitimate callings of life" ("Boone County Report," 1858–1859, p. 10).

Advocates of adding entertaining features to the agricultural fair had the pulse of the rural community. There was little sense in fighting a losing battle against the horse race and the popularization of the agricultural fair. Although it had started out (primarily) as forum for wealthy farmers to display their livestock, the agricultural fair had continually expanded its program and turned its attention inward toward the town. It was dismissed by some farmers as a town or city fair, but with the support of townsfolk, the annual agricultural fair had become the county fair. It had achieved a permanent place alongside Election Day and July Fourth as a premier civic event.

Assessing Farmers' Agricultural Learning and the Fair's Utility as an Educator

What, exactly, people were learning at the fair was an object of debate among leading agriculturists. In origin, agricultural fairs were supposed to be serious events, intended to encourage farmers to labor intelligently, to invest themselves in their calling. Broadened to encompass the town-based trades, the mission was to encourage all honest industry and useful arts; broadened still further to accommodate the ladies' flowers and needlework, to instill a desire for good taste and refinement. Regardless of the premium competition, the aim was to cultivate the useful and the good.

Little that was useful or good could be found at a horse race. Nor, thanks to the crowds that the races drew to the fairs, could these virtues could be at the county fair's gates, among the "bedlam clan" who pitched "their broad-road tents around our grounds, to gather dimes by corrupting morals" (Hobbs, 1859, p. 100). The auction stands, freak shows, gambling stalls, and hucksters were finding their way inside, sometimes without—

sometimes with—the blessing of agricultural societies. Little virtue could be claimed for the "baby shows" that were being passed off in some places as premium competitions, where "the contents of the cradle and the hog pen are judged by the same standard, when the babies are estimated by the pound, like fat calves in the shambles" ("Baby Shows," 1854, p. 49) Even less virtue was on display at one agricultural fair, where the audience found the "most attractive part" to be "a slave pen" in which "white children were sold at six dollars a head" ("The Fair at Attica," 1859, p. 2). Most serious agriculturists agreed that the agricultural fair had degenerated: a few solitary judges studied the pumpkins while the crowds gathered around the race course drinking whiskey and gambling ("A Modern Agricultural Fair," 1863). If the fair had anything to do with agricultural improvement, it was eclipsed by the aim of drawing and pleasing the crowds.

Since county fairs were the only institutions devoted to agricultural improvement in Indiana, Ignatius Brown, Secretary of the State Board of Agriculture, thought it was time to appraise their influence. One thing was plainly evident: "great changes" had "taken place within the last ten years in agriculture" (State Board of Agriculture, 1857b, p. 95). A comparison of the United States Census Bureau statistics for 1850 and 1860 confirms Brown's conclusion. Hoosier farmers added more than three million improved acres of farmland, tripling the value of their farms. The money invested in implements nearly doubled. Farmers owned more cattle, made more butter and cheese, obtained nearly as much wool from one-third fewer sheep. They grew more crops: market garden vegetables increased 650 percent; orchard products' value almost tripled; rye and barley increased 748 percent and 486 percent. Indiana's soil yielded more flax, sorghum, potatoes, and other crop varieties than it had a decade earlier. Some Hoosier farmers were improving their farms and abandoning exclusive reliance on the corn-and-hog-combination; of that there was no doubt.

But how much of the change in agriculture should be attributed to agricultural fairs? No one could deny that fairs had succeeded admirably "in directing public attention to the subject of agricultural improvement." The competitions and displays "excited a spirit of emulation" among farmers "when other agencies might have failed." But was it credible to maintain that an event "so transient and local" as a fair held only a few days each year could "produce such broad and lasting" results? The claims of fair-boosters notwithstanding, Ignatius Brown thought not: "some agency more powerful, permanent and constant in its effects" was responsible. Far more influential "than any number of fairs," the railroad was teaching farmers to improve (State Board of Agriculture, 1857b, pp. 98–100).

Prior to the coming of the railroad, many farmers seemed to have forgotten "that there is in the English language the word 'improvement.'" With access opened to the world's markets, they were "applying it to everything connected with agriculture" ("Knox County Report," 1854–1855, p. 79). The railroad made it easier to transport bulky grain crops and livestock long distances and to receive higher prices for them. It supplied the means to improve in the form of imported livestock and farm implements. Most important, it gave incentives in the form of consumer goods and modern conveniences. Between 1850 and 1860, as Indiana's population and nearly every economic sector grew rapidly, homemade manufactures declined 40 percent. With each household item that he purchased, the farmer found reason to improve his cultivation, livestock, and farm management. The conclusion to be drawn from this was stark. The railroad had compelled farmers "to make rapid improvements in their animals and products, and they must have made them though no societies had existed and no fairs had been held" (State Board of Agriculture, 1857b, p. 100).

For Ignatius Brown and others, the nature of the agricultural fair as an educational institution was questionable. What were farmers learning? If nine-tenths of those who attended did so out of "curiosity or a desire for recreation," could one feel confident that they had learned anything of value? The "perpetuation of the information" exhibited at fairs could not be assured. From one year to the next, most of the people perceived "no marked difference" in the livestock, crops, and machinery displayed. Neither the judges nor the audience were able "to retain clear ideas of the good points of the animals or articles exhibited." No matter how much awe the fair's exhibits might evoke among visitors on the grounds, they were not inspiring (enough) farmers to investigate improved agricultural methods throughout the year (State Board of Agriculture, 1857b, pp. 100–102).

Equally troubling, attendance at fairs was not conveying to farmers the full lessons of agricultural improvement. Indiana's farmers were willing to adopt technological innovations and new crops. But, notwithstanding the decade-long campaign to promote a variety of alternative (or supplemental) agricultural products, the importance of traditional crop staples continued to grow. Again, comparison of United States Census Bureau statistics from 1850 and 1860 reveal the trend. Wheat cultivation doubled, corn remained king, and Hoosier hogs still outnumbered Hoosier people by 2.3 to 1. Most farmers continued to farm extensively, tilling large acreage with little soil renewal, rather than intensively, using the techniques of diversified scientific agriculture. Agricultural improvers' prescriptions were no match against high prices for staple crops, cheap land values, and improved horse-drawn implements. Replacing inferior stock with mixed breed and purebred animals and using implements was how most farmers improved their commercial farming practices. Only a small number of farmers, those who lived in the immediate vicinity of cities and villages, were diversifying their market basket to supply the needs of town-dwellers.

With these sorts of things in mind (among others), Secretary Ignatius Brown rendered his judgment upon the agricultural fair's utility as an educator: "The day has probably gone by, if it ever existed—when the mere dispensing of premiums could effect lasting improvements" in agriculture (State Board of Agriculture, 1857b, p. 100). So long as fairs remained the only institutions devoted to agricultural education in Indiana, customary practice and farmers' perception of immediate moneymaking prospects would determine which crops, livestock, and methods they adopted. A thorough understanding of agriculture's science and economy could not be disseminated through fairs alone. To advance agricultural knowledge, focused efforts—carried out by institutions capable of exerting a constant influence upon the minds of leading farmers—were required. A combination of agricultural colleges, model farms, and experimental premiums would have to be established (State Board of Agriculture, 1857b, p. 16). With appropriate encouragement at the county fairs, common farmers would take advantage of such efforts (or not) as self-interest and changes in the conditions of life compelled them. This was the lesson taught by the railroad and nearly a decade's worth of educational outreach frustration.

To diffuse correct ideas and techniques among the farmers, the fairs would have to rediscover their original strictly agricultural mission. To the dismay of agriculturists, the town-boosting mission took the county fair in the opposite direction and improving farmers turned to other agencies for their agricultural education. No longer much of an agricultural fair, the county fair continued to serve as a prominent source of farmers' civic education. Most farmers had taken little notice of the civic aspirations attached to the fair during the 1850s. After the Civil War, they became very aware of its civic mission to bring the loom and the anvil into proximity with the plow. Most were not pleased with what that represented or with how townsmen were using supposedly agricultural fairs to achieve it. Occupying a visible place in the public eye—and directly concerned with economic development—the county fair was becoming a flashpoint for controversies dividing the town from the country.

Reflections on Civic Learning

In the antebellum era, the northern states witnessed the flourishing of a variety of reform movements. Among them, the movements for slavery's abolition, temperance, common schools, and women's rights are the most famous. Indiana's agricultural improvement campaign of the 1850s shared features with them, and social improvements of various kinds were expected to follow from it. As with any movement, it was intended to organize people of like minds into a common venture and to bring other people into the fold. The agricultural improvement campaign centered squarely on the economic good it might yield for the community, and individuals' self-discerned best interest was upheld as its single most important driving force. The men involved hoped that the power of self-interest would draw farmers into town to learn about agricultural improvement's benefits through monthly experience meetings and annual fairs.

The agricultural societies were not farmers' organizations, strictly speaking. Instead, in each county, they brought a few dozen leading farmers and townsmen into a coalition. Townsmen, naturally, were more interested in economic development and fair-hosting, but they recognized that these aims could not be achieved unless farmers' improvement ambitions were stimulated. Farmers who were involved in agricultural societies shared these goals. But, they were not typical farmers. They were gentlemen farmers who had retired from town-based professions to rural estates and the very best farmers, men who achieved considerable wealth from a lifetime of farming and other pursuits. The consequence of this blending of townsmen and wealthy farmers was that agricultural societies were stamped with a town character. As such, the agricultural improvement campaign was not all that different from the passage of the Mammoth improvements projects in the 1830s. It, too, was initiated by the most influential actors of the rural community for the good of the farmers and the people generally.

The leading farmers who joined agricultural societies in the 1850s expected their example and instruction to be heeded. Who better to teach farmers how to take advantage of the emerging marketing opportunities created by local economic growth and closer commercial ties with the Northeast? Most, but not all, farmers rejected this claim to leadership and the comprehensive approach to agricultural improvement that accompanied it. With labor scarce and crop staple prices high, they saw little reason to engage in the intensive farming that put vegetables, fruits, and dairy products on town-dwellers' tables. Farmers' self-interest, as they conceived it, did not lead directly toward the behavior that, on agricultural improvers' account, would have been most beneficial to the community as a whole.

Neither agricultural societies nor fairs were well calculated to close the gaps in personal experience, education, wealth, ambition, and civic concern between leading farmers and common farmers. The typical farmer could not really participate in these institutions as they were conducted in the early 1850s. At best, he could attend agricultural fairs to have a good time, to admire the livestock, and to learn about potential ways to improve his farm from conversing with other men who found themselves in situations similar to his own. As a civic event that brought rural people into contact with each other, the county fair provided the opportunity for some agricultural learning. The broader civic learning it offered, however, may have been more instrumental in changing farmers' behavior. When surrounded by the county fair's displays and activities, no one could deny that profound changes were underway. Imported fancy livestock, new types of farm machinery, and consumer goods testified to it, as did the appearance of horse races and gambling stalls.

Ultimately, the significance of the agricultural education campaign of the 1850s may be tied less to the farmers' improvement than to what it reveals about the leadership of the rural community. With their prescriptions for agricultural improvement rejected, the prominent farmers of Indiana's counties did not abandon their claim to being the rural community's best men. Nor did they relinquish their conviction that they were the best judges of what farmers ought to do as producers for the commercial economy. However, they withdrew from direct efforts at educational outreach and turned over the agricultural societies to townsmen's control. In doing so, prominent farmers did not quite cede leadership of the rural community to men of the towns. They agreed, more or less, on a common agenda, after all, and there were other leadership outlets. Nevertheless, their relinquishment of the agricultural societies was a telling sign. Farmers had always set the tone in Indiana, but the institutions and imperatives of towns were coming into their own as the driving forces of change.

References

Act authorizing county agricultural societies to purchase and hold real estate. (1855). *Indiana Laws* (38th session), 49.

Act authorizing the State Board of Agriculture. (1859). *Indiana Laws* (40th session), 112–113.

Agricultural meeting. (1852). *Indiana Farmer, 1*(14), 210.

Agricultural spirit. (1851). *Indiana Farmer, 1*(8), 123.

Allen County agricultural society—list of premiums. (1855, October 18). *Fort Wayne Daily Times.*

Baby shows. (1854). *Indiana Farmer, 4*(4), 49–50.

Barnes, W. A. (1852). Address to the Porter County fair. In *Annual Report of the Indiana State Board of Agriculture.* Indianapolis, IN: State Board of Agriculture.

Barnhart, J. D., & Carmony, D. F. (1954). *Indiana: From frontier to industrial commonwealth* (Vol. 2). New York: Lewis Historical Publishing Company.

Bartholomew County report. (1853). In *Annual Report of the Indiana State Board of Agriculture.* Indianapolis, IN: State Board of Agriculture.

Bartholomew County report. (1857). In *Annual Report of the Indiana State Board of Agriculture.* Indianapolis, IN: State Board of Agriculture.

Bartholomew, H. S. K. (1930). *Pioneer history of Elkhart County, Indiana.* Goshen, IN: The Goshen Printery.

Blanke, D. (2000). *Sowing the American dream.* Athens: Ohio University Press.

Boone County report. (1852). In *Annual Report of the Indiana State Board of Agriculture.* Indianapolis, IN: State Board of Agriculture.

Boone County report. (1856). In *Annual Report of the Indiana State Board of Agriculture.* Indianapolis, IN: State Board of Agriculture.

Boone County report. (1857). In *Annual Report of the Indiana State Board of Agriculture.* Indianapolis, IN: State Board of Agriculture.

Boone County report. (1858–1859). In *Annual Report of the Indiana State Board of Agriculture.* Indianapolis, IN: State Board of Agriculture.

County fair. (1856, August 27). *Wabash County Weekly Intelligencer,* p. 2.

Dearborn County report. (1854–1855). In *Annual Report of the Indiana State Board of Agriculture.* Indianapolis, IN: State Board of Agriculture.

Dearborn County report. (1858). In *Annual Report of the Indiana State Board of Agriculture.* Indianapolis, IN: State Board of Agriculture.

Dearborn County report. (1859). In *Annual Report of the Indiana State Board of Agriculture.* Indianapolis, IN: State Board of Agriculture.

Decatur County report. (1852). In *Annual Report of the Indiana State Board of Agriculture.* Indianapolis, IN: State Board of Agriculture.

Dekalb County fair. (1859, October 24). *Wabash Plain Dealer,* p. 2.

Delaware County report. (1854–1855). *Annual Report of the Indiana State Board of Agriculture.* Indianapolis, IN: State Board of Agriculture.

Elkhart County report. (1851). In *Annual Report of the Indiana State Board of Agriculture.* Indianapolis, IN: State Board of Agriculture.

Elkhart County report. (1851–1852). In *Annual Report of the Indiana State Board of Agriculture.* Indianapolis, IN: State Board of Agriculture.

The fair at Attica. (1859, September 21). *The People's Friend,* p. 2.

Fancy farmers. (1852). *Indiana Farmer, 1*(24), 379–380.

Fayette County report. (1853). In *Annual Report of the Indiana State Board of Agriculture.* Indianapolis, IN: State Board of Agriculture.

Fayette County report. (1854). In *Annual Report of the Indiana State Board of Agriculture.* Indianapolis, IN: State Board of Agriculture.

Fayette County report. (1854–1855). In *Annual Report of the Indiana State Board of Agriculture.* Indianapolis, IN: State Board of Agriculture.

Floyd County report. (1858). In *Annual Report of the Indiana State Board of Agriculture.* Indianapolis, IN: State Board of Agriculture.

Franklin County report. (1852). In *Annual Report of the Indiana State Board of Agriculture*. Indianapolis, IN: State Board of Agriculture.

Franklin County report. (1853). In *Annual Report of the Indiana State Board of Agriculture*. Indianapolis, IN: State Board of Agriculture.

Franklin County report. (1854–1855). In *Annual Report of the Indiana State Board of Agriculture*. Indianapolis, IN: State Board of Agriculture.

Gates, P. (1960). *The farmer's age: Agriculture, 1815–1860*. New York: Holt, Rinehart and Winston.

Hamilton County report. (1858). In *Annual Report of the Indiana State Board of Agriculture*. Indianapolis, IN: State Board of Agriculture.

Hendricks County report. (1853). In *Annual Report of the Indiana State Board of Agriculture*. Indianapolis, IN: State Board of Agriculture.

Hendricks County report. (1854–1855). In *Annual Report of the Indiana State Board of Agriculture*. Indianapolis, IN: State Board of Agriculture.

Hendricks County report. (1858). In *Annual Report of the Indiana State Board of Agriculture*. Indianapolis, IN: State Board of Agriculture.

History of Greene and Sullivan Counties, State of Indiana. (1884). Chicago: Goodspeed Brothers.

History of St. Joseph County, Indiana. (1880). Chicago: Charles C. Chapman and Company.

Hobbs, B. C. (1859). Address to the Parke and Vermillion agricultural fair. In *Annual Report of the Indiana State Board of Agriculture*. Indianapolis, IN: State Board of Agriculture.

Huntington County report. (1857). In *Annual Report of the Indiana State Board of Agriculture*. Indianapolis, IN: State Board of Agriculture.

Innis, G. S. (1859). Equalizing the premium lists. *Ohio Cultivator, 15*(1), 6.

Knox County report. (1854–1855). In *Annual Report of the Indiana State Board of Agriculture*. Indianapolis, IN: State Board of Agriculture.

Kosciusko County report. (1857). In *Annual Report of the Indiana State Board of Agriculture*. Indianapolis, IN: State Board of Agriculture.

Lagrange County report. (1856). In *Annual Report of the Indiana State Board of Agriculture*. Indianapolis, IN: State Board of Agriculture.

LaPorte County report. (1858–1859). In *Annual Report of the Indiana State Board of Agriculture*. Indianapolis, IN: State Board of Agriculture.

LaPorte County report. (1859). In *Annual Report of the Indiana State Board of Agriculture*. Indianapolis, IN: State Board of Agriculture.

Madison County report. (1856). In *Annual Report of the Indiana State Board of Agriculture*. Indianapolis, IN: State Board of Agriculture.

Mahoney, T. R. (1990). *River towns in the great west: The structure of provincial urbanization in the American Midwest, 1820–1870*. New York: Cambridge University Press.

Marion County report. (1854–1855). In *Annual Report of the Indiana State Board of Agriculture*. Indianapolis, IN: State Board of Agriculture.

Marshall County report. (1859). In *Annual Report of the Indiana State Board of Agriculture*. Indianapolis, IN: State Board of Agriculture.

Miami County report. (1857). In *Annual Report of the Indiana State Board of Agriculture*. Indianapolis, IN: State Board of Agriculture.

A modern agricultural fair. (1863). *American Agriculturist, 22*(11), 329.

Monroe County report. (1857). In *Annual Report of the Indiana State Board of Agriculture*. Indianapolis, IN: State Board of Agriculture.

Morgan County report. (1856). In *Annual Report of the Indiana State Board of Agriculture*. Indianapolis, IN: State Board of Agriculture.

Ohio and Switzerland County district report. (1853). In *Annual Report of the Indiana State Board of Agriculture*. Indianapolis, IN: State Board of Agriculture.

Porter County report. (1853). In *Annual Report of the Indiana State Board of Agriculture*. Indianapolis, IN: State Board of Agriculture.

Posey County report. (1859). In *Annual Report of the Indiana State Board of Agriculture*. Indianapolis, IN: State Board of Agriculture.

Putnam County report. (1852). In *Annual Report of the Indiana State Board of Agriculture*. Indianapolis, IN: State Board of Agriculture.

Rush County report. (1853). In *Annual Report of the Indiana State Board of Agriculture*. Indianapolis, IN: State Board of Agriculture.

Rush County report. (1856). In *Annual Report of the Indiana State Board of Agriculture*. Indianapolis, IN: State Board of Agriculture.

Shelby County report. (1854–1855). In *Annual Report of the Indiana State Board of Agriculture*. Indianapolis, IN: State Board of Agriculture.

St. Joseph County report. (1857). In *Annual Report of the Indiana State Board of Agriculture*. Indianapolis, IN: State Board of Agriculture.

State Board of Agriculture. (1857a). [Table of premiums awarded at fairs]. In *Annual Report of the Indiana State Board of Agriculture* (pp. 713–714). Indianapolis, IN: Author.

State Board of Agriculture. (1857b). Secretary's report. In *Annual Report of the Indiana State Board of Agriculture*. Indianapolis, IN: Author.

State Board of Agriculture. (1858). Proceedings. In *Annual Report of the Indiana State Board of Agriculture*. Indianapolis, IN: Author.

State Board of Agriculture. (1858–1859). Proceedings. In *Annual Report of the Indiana State Board of Agriculture*. Indianapolis, IN: Author.

State Board of Agriculture. (1859). Preface. In *Annual Report of the Indiana State Board of Agriculture*. Indianapolis, IN: Author.

Switzerland and Ohio district report. (1859). In *Annual Report of the Indiana State Board of Agriculture*. Indianapolis, IN: State Board of Agriculture.

Tippecanoe County report. (1851). In *Annual Report of the Indiana State Board of Agriculture*. Indianapolis, IN: State Board of Agriculture.

Vigo County report. (1856). In *Annual Report of the Indiana State Board of Agriculture*. Indianapolis, IN: State Board of Agriculture.

Warren County report. (1854–1855). In *Annual Report of the Indiana State Board of Agriculture*. Indianapolis, IN: State Board of Agriculture.

Warren and Fountain district report. (1856). In *Annual Report of the Indiana State Board of Agriculture*. Indianapolis, IN: State Board of Agriculture.

Warrick County report. (1858–1859). In *Annual Report of the Indiana State Board of Agriculture*. Indianapolis, IN: State Board of Agriculture.

Washington County report. (1857). In *Annual Report of the Indiana State Board of Agriculture*. Indianapolis, IN: State Board of Agriculture.

Washington and Orange district report. (1853). In *Annual Report of the Indiana State Board of Agriculture*. Indianapolis, IN: State Board of Agriculture.

Weir, J. (1852). [Untitled letter to the editor]. *Indiana Farmer, 1*(10), 152.

Women riding at the fairs. (1855). *Indiana Farmer,* 4(7), 98–99.

Wright, J. A. (1851a). Address to the Wayne County agricultural fair. In *Annual Report of the Indiana State Board of Agriculture*. Indianapolis, IN: State Board of Agriculture.

Wright, J. A. (1851b). President's report. In *Annual Report of the Indiana State Board of Agriculture*. Indianapolis, IN: State Board of Agriculture.

Wright, J. A. (1852). President's report. In *Annual Report of the Indiana State Board of Agriculture*. Indianapolis, IN: State Board of Agriculture.

Wright, J. A. (1853). Address to the Washington and Orange Agricultural Fair. In *Annual Report of the Indiana State Board of Agriculture*. Indianapolis, IN: State Board of Agriculture.

4

Growing Indiana

Agricultural Improvement and the Growth Imperative

The Impact of the Civil War on Economic Development Ambitions in Indiana

With the southern states' secession, agricultural improvement took a back-seat to the immediate necessities of war. Divided bitterly in their politics throughout the 1850s, Hoosiers joined citizens throughout the North in rallying to the flag as word of the attack on Fort Sumter swept the country-side. A flurry of mass meetings and a rush to volunteer for military service, dispelled doubts as to whether Indiana—with her close commercial and fraternal ties to southern states—would remain loyal to the Union. Patriotic solidarity waned quickly. Old fault lines in public opinion reformed around war issues. Should the war be largely defensive, to preserve the Union as it was, or should it be used to end slavery? What state and national govern-ment measures were justifiable on the grounds of war necessity? To what ex-tent should citizens sacrifice their liberties and lend their mite to the mili-tary campaign? These big questions were raised everywhere. In no northern state were they asked and answered more divisively than in Indiana.

Civic Learning through Agricultural Improvement, pages 73–95
Copyright © 2011 by Information Age Publishing
73

War aims and war measures occupied the press and politics. Scant support for waging war to end slavery could be found in Indiana. President Lincoln's issuance of a preliminary Emancipation Proclamation in the summer of 1862 gave Indiana Democrats sweeping victories at the autumn polls. Policies implemented by the Republican administration corroded Hoosier support for the war effort: protective tariffs aided Eastern manufacturers, paper currency Eastern bankers; both, when combined with profiteering, artificially inflated prices for supplies. Repeated Union losses in battle stoked criticism of the war's conduct. The heavy-handed measures used to suppress treasonous activities (alleged and genuine) and to organize Indiana's campaign by the state's governor, Oliver Morton, generated still more. Accusations of disloyal conspiracy (against Democrats) and of scare mongering (against Republicans) filled the press. The state government broke down when Republicans bolted the 1863 legislative session and Democrats, in turn, blocked Governor Morton's access to public funds. Copperhead-ism and Union-ism shared close quarters, nowhere closer than among the neighbors who clashed at political rallies, among the mobs destroying newspaper offices and attacking men commissioned to implement the draft, and between the rival bands of armed citizenry, the Sons of Liberty and Union Clubs.

The effects of war consumed Hoosier social life. Over half of Indiana's men of military age enlisted for three-year terms in the Union Army, leaving women and children to plant crops, tend livestock, and manage farms. Thousands more joined volunteer groups to work behind the lines handling supplies, guarding depots, and carrying communications, or, in the case of the Indiana Legion, to defend against Confederate squadrons who crossed the Ohio River to steal horses, food, and supplies. The war stretched from the battlefield into home communities as the wounded returned from previously unknown places like Antietam, Chancellorsville, Shiloh, and Chickamauga. New recruits and draftees, nurses, and agents for commission and supply went in the other direction. Accompanying them was the money and goods—Bibles and bedding, food and clothing, paper and medical supplies—solicited by soldiers' aid societies, church groups, and the auxiliary societies of the Indiana Sanitary Commission that sprang up in most counties. Supplying the troops, caring for veterans and their dependents, struggling to survive in the face of high prices and constant anxieties formed the bulk of Hoosier priorities (Thornbrough, 1965).

Using arguments of wartime necessity to brush aside constitutional scruples and western objections, Congressional majorities that were overwhelmingly Eastern and Republican, overhauled federal economic policy. Some of the policies were very popular in Indiana and other western states.

Having been transplanted two and three times (or more), most westerners favored legislation that would make it easier for settlers to make farms and harder for wealthy speculators to lock up the land against the cultivator. Westerners had never been wanting in enthusiasm for building railroads; laying new lines across the Great Plains met with their hearty approval. Other policies were accepted grudgingly, at best, or opposed strenuously. A comprehensive protective tariff program, inflated currency (through issuing greenbacks and treasury notes), and a national banking system met hostile reactions. These changes were justifiable on the basis of the war effort. It was equally clear to Indiana Democrats that they coincided with the agenda that the Republicans had championed since the party's origin, and the Whigs before them. When the war ended there was little doubt that federal policy gave decisive advantages to Eastern interests while stifling western efforts at economic development (Richardson, 1997).

Changes in financial policy had the most striking effect. To finance the war, the United States government borrowed heavily, amassing a debt burden of $2.6 billion. Borrowed at high rates of interest, the debt helped to increase eastern capitalists' wealth and influence over economic policy. In the short run it led to the creation of a national banking system that (through further wartime legislation) squeezed state bank notes out of circulation. The preponderance of reserves was assigned to eastern banks. This favoritism (as Midwesterners charged) increased westerners' dependence upon eastern credit and financial decisions. Their dependence was compounded by the federal government's issuing more than $700 million in greenbacks and treasury notes that were not backed by gold. Wealthy investors who bought devalued currency during the war demanded repayment in gold and a contraction of the money supply after the war ended. Joined by eastern merchants (who needed hard currency to import goods), their lobbying succeeded. The net result of the changes in financial policy was the drawing away of western capital to the eastern seaboard (Donald, Baker, & Holt, 2001).

Fear of increased dependency, in part, was behind traditional Hoosier opposition to paper currency. Hard currency was sound, less susceptible to manipulation. The same fear had inspired opposition to national banking and support for state banks and free banking. Locally issued bank notes were, to a degree, extensions of traditional borrowing and exchange among people who shared common interests. They also helped to ensure that wealth generated within a community remained in local circulation. The change in fiscal policy ensured that wealth flowed out. This result was compounded by the effects of the protective tariff program. Passed to raise revenue and to encourage production, it placed high duties on a wide range of agricultural and manufactured products. Agriculture gained nothing from tariff protection; sheltered from foreign rivals and called on to meet wartime demand, the Northeast's manufacturing industries gained a decisive head start on nascent Midwestern industries in scale, efficiency, and in creating networks of supply and distribution. Together, the revised fiscal and tariff policies conspired, as Thomas A. Hendricks warned Indiana's Democratic convention in 1862, to make westerners "the 'hewers of wood and drawers of water' for the capitalists of New England and Pennsylvania." Unsurprisingly, the policies were denounced at mass meetings throughout Indiana (Thornbrough, 1965, pp. 133–134).

If ever the call for promoting "home interests" resonated among Indiana's townsmen, it was among the post-Civil War generation. Was not it folly for Hoosiers to be "ruinously tributary to the East" while "mines of wealth" rivaling the silver of Peru lay buried under their feet? Who could be so "blind and dull" as to fail to see why the Northeastern prospered? Or, that the Western states struggled because they had "confined their attention almost wholly to agricultural pursuits?" There was no point in condemning easterners. The appropriate remedy was to "strive to emulate their example." The main barrier to doing so was the absence of investment capital. Hoosiers would need to attract it by establishing "the fact before the capitalists that [Indiana] is as rich a field for manufacturing enterprise as any to be found, while the [railroad] facilities are full as good, and the incidental circumstances [location] better" (Bland, 1867, pp. 156, 174). Extraordinary efforts were needed to generate the full potential of Indiana's resources and advantages, to empower its communities to become self-sufficient, that is, to be able to meet their citizens' needs and to trade the surplus with the world on favorable terms. The problem and solution comprised the same agenda that Governor Joseph Wright had identified during the 1850s. Intensified by the impact of the war and the transformation of federal policy, Wright's doctrine of bringing the loom and the anvil into proximity with the plow took on the nature of a civic imperative.

Indiana's development aims and sense of urgency were hardly unique. They were shared by community leaders throughout the Midwest. Historical memory of the post-Civil War economy conjures up an image of consolidation and concentration, the expansion of economic control across geographic space and the gathering of resources into a smaller number of places and firms. Consolidation began, however, in the late 1850s, as the building of local railroad lines created a host of overlapping market- and supply-zones that, taken together, brought the Midwest into a common market with the Northeast. The immediate post-Civil War years witnessed the proliferation of competitive forces. People suddenly found themselves very aware of the economic activity that occurred in other places. Larger towns that possessed railroad lines and significant industries sought to build on their existing advantages. Smaller towns that had been bypassed by railroad lines sought to create them, as well as to attract manufacturing enterprises and the investment dollars needed to grow. Eager to be of service, state governments (on the grounds of state self-sufficiency) empowered localities to lend public credit to entrepreneurs as well as to raise special taxes for infrastructure and development purposes. The result was a mad scramble between states and between communities for the means of promoting economic growth. The great wave of concentration began after the post-war boom ended with the financial panic of 1873 (Donald et al., 2001).

Heeding the Growth Imperative

In Indiana, the development drive was expressed most prominently in railroad-building. A score of new trunk lines were built to connect major cities to each other and the world; feeder lines were built to them from lesser cities. The coalfields of the southwest region were opened with several more lines. Indiana's General Assembly encouraged these efforts by empowering counties to raise funds through taxation, if citizens approved referenda. Public aid for manufacturing, however, did not receive the state's blessing. A bill on behalf of manufacturing came close to passage in 1867, while Republicans held power. All similar proposals in the following years were blocked by Democratic and farmer-backed legislators. Farmers supported railroad building because it gave them access to the markets and goods of the East. Toward the presence (or lack) of manufacturing in nearby towns, farmers were largely indifferent. Toward the encouragement of manufacturing's growth through public subsidy (using their tax dollars), most farmers were stoutly opposed. Deprived of the opportunity to use public aid to jump start manufacturing, Indiana's town boosters were forced to rely on public relations campaigns and private voluntary efforts.

Capital and labor to develop the untapped natural resources of their communities were what the leaders of Indiana's communities wanted. Why would investment-seeking capitalists and employment-seeking workers make their way to Indiana when they knew next to nothing about the state? Illinois, Missouri, Michigan, Wisconsin, more distant and younger states further west, were better-known. Indiana's government had done almost nothing to publicize the state. It had commissioned few promotional pamphlets; it sent no agents to the East to convince aspiring entrepreneurs to relocate. It refused to follow the lead of the United States (as other states were doing) in establishing a Board of Immigration for enticing new migrants from Europe. The only promotional work of significance that Indiana had done was to commission a geological survey. Published during the Civil War, when no one was paying attention, the survey's results did little to aid economic development. After the war, the survey resumed, but Indiana's General Assembly allocated far less than other states. A yearly appropriation of $5,000 was better than nothing, but (according to its advocates) at least $20,000 was needed to survey the state's resources properly and to disseminate the information effectively. Other promotional outlets would have to supplement the geological survey (State Board of Agriculture, 1867b, pp. 224–225).

At the agricultural improvement campaign's outset in the 1850s, it was expected that the State Board of Agriculture's *Annual Report* would serve primarily as a means of gathering and exchanging agricultural information. An expensive volume with a limited circulation of 3,000 copies, the *Annual Report* was intended for members of the agricultural societies and the General Assembly. Hoping to make it a useful reference manual, the State Board required each county society to prepare a detailed report on local agriculture (based on a questionnaire). The quality of the county reports fell short of expectations, but the *Annual Report* was filled with accounts of fair-hosting efforts, descriptions of growing conditions, and a smattering of statistical data. Estimating the publication cost to be about $9,000 the General Assembly—without a dissenting vote—declared the sum to be "more than the Reports are worth" and discontinued publication in 1861 ("A Joint Resolution," 1861, p. 183). With some difficulty, the State Board of Agriculture managed to get the financing restored after the Civil War.

Under the influence of the growth imperative, the *Annual Report* became a promotional volume. "Frequent calls" for information about Indiana's prospects were received from the eastern states, the State Board's Secretary informed members of county societies (and the General Assembly). If reports were filled with "accounts of the mineral and agricultural resources of the State, their beneficial effect... in inducing emigration to the State, could hardly be over-estimated" (State Board of Agriculture, 1867b, p. 7). The secretaries of the county agricultural societies (who seldom were farmers) began extolling the virtues of their counties: climate and location, timber and ores, rail facilities and manufacturing establishments, the people's character and industry, and the quality of public schools. The *Annual Report* reached its peak of perfection as promotional literature in 1876, when every county was profiled and a full complement of articles (and maps) on Indiana's natural resources was compressed between its covers. Precisely what effects this information had in enticing capital and labor to Indiana is unknowable, but the General Assembly was convinced of its utility as early as 1872. For the next year, 10,000 copies were funded for the purpose of distributing them to agricultural societies and libraries in other states (State Board of Agriculture, 1873, p. 336).

Few of Indiana's farmers read the State Board of Agriculture's *Annual Report.* Even if it had been offered to them, why would they? Concern for including agricultural substance dropped markedly in the 1870s, after the State Board abandoned its requirement that county societies respond to its agricultural questionnaire (State Board of Agriculture, 1868, p. 64). Extolling Indiana's "wonderful resources and advantages" and its "geographical position and favored locality" took precedence. The state had ample resources and advantages to exploit. Above all, though, Indiana had plenty of room to grow. It had, on average, only 50 persons to the square mile. Other portions of the globe supported a population density of 400 persons to the square mile. Why could not—and should not—the population of 2 million be increased to 17 million? (State Board of Agriculture, 1875b, p. 6). In the 1870s, this sort of thinking gave rise to rabid boosterism, promising unheralded prosperity, limitless growth, and the overnight transformation of sleepy market towns into great commercial centers. It was irrational—a far cry from the slow, but steady prescription of true progress—but it was not wholly unfounded.

The economic boom engendered by wartime demand fueled Indiana's town-boosting ambitions. If the decade between the Mexican-American War and the 1857 financial panic had been the first golden age for Indiana's farmers, the 1860s was the first for her manufacturers. The old export standards—flour, meat, and lumber—continued to grow. The dynamism was not in processing the raw materials of farm and forest, however, but in making finished products. According to United States Census Bureau estimates, between 1860 and 1870, the number of manufacturing establishments and employees more than doubled. High value-added items made from wood, such as carriages, wagons, and furniture, tripled in value. The textile industry saw a fourfold increase in value, as did the production of various kinds of machinery. Railroad car production had been unknown in Indiana during the 1850s. In the next decade, 10 large establishments went into operation; employing more than 1,400 workers, it was the 13th ranked manufacturing industry by 1870. Impressive though these gains were, they scarcely compared to the tremendous growth of the iron and steel industries. Ranked 16th among Indiana's manufacturing industries in 1860, products of the forge surged to 3rd rank by 1870 with a thirteen-fold increase in value. Between the growth of manufacturing and their role as exchange points, the populations doubled or tripled in towns that were situated favorably. On the whole, the state's population increased 25 percent during the 1860s (Barnhart & Carmony, 1954, pp. 237–240).

Places that had been mere villages on the eve of the Civil War were booming cities by the 1870s. South Bend was Indiana's most striking example, the home of the Studebaker Wagon Works, the Oliver Chilled Plow Works, the Birdsall Clover Huller factory, several large paper mills and ironworks, and dozens of smaller manufacturing establishments. Not far away, in Cass County, a similar scene could be witnessed. Booming as a result of the Chicago and Eastern Railroad Corporation's decision to locate foundries there, the city of Logansport drummed-up a $50,000 package of incentives to get the company to expand operations. Leading men in the adjacent county took notice. Employing similar means, they outbid all rivals to attract the Howe Sewing Machine Company; the Howe factory soon joined the state's largest woolens factory at the town of Peru. Hundreds of workers accompanied major projects and the supporting industries needed to supply raw materials. Tremendous growth was possible in places chosen as investment sites by people with capital and the spirit of enterprise (Thornbrough, 1965).

What did the fortunate places have that other places lacked? With the exception of the coal-producing counties in Indiana's southwestern region, few rural counties could claim much by way of peculiar natural advantages. Advances in steam engine technology made obsolete falling water as a requirement for manufacturing. With railroad lines stretching through every part of Indiana, virtually every county could claim prime access to world markets. Most counties (save the deforested southern counties and a few plains counties) had great stands of timber, both the prized hardwoods for furniture and carriages (especially black walnut, hickory, and oak) and the lesser woods suited for building materials and fuel. Every county had an abundance of cheap and fertile land, as well as an ample number of men who imagined that great lodes of ores and minerals were under the topsoil.

Indiana's cities faced the same problem as its counties. They were not all that different from each other. Indianapolis stood alone in preeminence, more than double the size of the next largest city by 1870; with gusto, boosters claimed Chicago and St. Louis as rivals. The next tier of cities (Evansville, Fort Wayne, Terre Haute, New Albany, Lafayette, and Madison), as regional centers, were evenly matched rivals, but they were considerably larger than all others. The rest of Indiana's county seats and market towns were small; their thousands could be counted on one hand, typically with a finger or two to spare. A city's ability to claim some mark of distinction was, to a considerable degree, tied to its surrounding county's location, resources, and people. Circumstance and chance—natural advantages and accidents of history—were the primary determinants of some important factors of economic growth, such as possessing mineral wealth or being located on a railroad line between major urban centers. Other factors of growth, particularly the presence of large scale manufacturing enterprises, could be artificially and deliberately cultivated. Developing a reputation as a center of growth and plying on the incentives to make that happen was the main thing that Indiana's cities and counties could do to set themselves apart from each other.

Attracting notice through high profile publicity was one way that Indiana's towns could respond to the growth imperative. That was the civic mission that supplanted agricultural education as the rationale for hosting fairs in the 1870s. County fairs had demonstrated some value in promoting economic activity in the late 1850s. Thriving fairs generated economic momentum by bringing farmers into town for a few days of revelry and by transferring their surplus capital into the hands of the town's leading businessmen. (Doing one's part for the public good had its rewards.) In addition to this role, in the post-Civil War years, with the need to attract the ingredients of growth from abroad all the more pressing, the annual fair became a feature event to put the resources, character, and energy of a county (or city or state) on display before an audience of visitors from other places. The "chief object of fairs" was "to make the best show possible" of the "special advantages" of a community. People who fulfilled this mandate pursued a "wide-awake course" ("County Fairs," 1874, p. 4). Leading townsmen were certain that they, their town, and their county would benefit from hosting a successful fair.

Undeniably, there was something to the notion. Big fairs brought high-rollers and representatives from major corporations into Indiana's cities. If they came primarily to promote their own fortunes, so be it. They might also be induced to loan capital to promising local talent, or, better yet, to invest in a local branch office, warehouse, or workshop. Whether through the fair's displays, behind-the-scenes meetings with local citizens, or tours of the city and county, wealthy visitors from abroad had the opportunity to take stock of what the fair-hosting place might have to offer. The same could be said for the less wealthy, but no less aspiring, residents of other counties in Indiana who might consider relocating for a new farm or business opportunity. How many people chose this particular town or that one because of what they saw and who they met on a visit to a fair is unknowable. Perhaps it was not so much the fair itself as it was the opportunity for making social connections that made the difference? Perhaps name recognition was enough? Cities then, as now, have done far more than host fairs for that elusive quality. Ultimately, it does not matter how fairs worked to boost cities. People were convinced they did. A mania for hosting fairs swept Indiana's leading townsmen as they imagined what a fair might do for their fortunes if its potential as an engine of economic growth was fully realized.

Fair-Hosting for Boosting the Town Economy

In the midst of the Civil War, Hoosiers had little use for agricultural fairs. Most were suspended, replaced by sanitary fairs and fund- and supply-raising events hosted by charitable organizations. Processions of farm wagons loaded with grains, vegetables, and firewood replaced horse races as the feature attraction. Joined by brass bands and marching soldiers, they displayed people's willingness to contribute to soldiers' families. Fairgrounds were used for these events as well as for musters and drills of local companies. (The state fairgrounds became known as Camp Morton, being used, in succession, as a militia bivouac, a prisoner of war camp, and a military hospital.) The few agricultural fairs held were given over largely to war-related purposes, and the proceeds donated to soldiers' aid societies.

Immediately following the war, crop failures, rainy weather, and "the want of a proper appreciation of the value of agricultural societies as 'institutions' upon the part of the agricultural community" checked revival efforts ("Marshall County Report," 1867, p. 470). Despite the setbacks, at least 21 fairs were hosted in the fall of 1866 ("District Fairs," 1866, p. 133). Townsmen who wanted to turn their fairs into showcase events discovered quickly that they had the means at hand. In its last action on behalf of agricultural improvement before the outbreak of war, Indiana's General Assembly—with nearly unanimous assent—had given its blessing to the emerging mission of boosting towns by hosting showcase fairs. It authorized agricultural societies to turn themselves into joint stock companies. Equipped with the right to borrow as much as $10,000 and to own 80 acres of land, the refashioned agricultural societies could expand operations by building bigger and better facilities ("Act to Amend an Act," 1861).

As joint stock associations, the county agricultural societies were no longer required to have directors from all of the townships within their counties' boundaries (as stipulated in the organic act of 1851). Instead, since the resources of any people willing to purchase shares could be pooled, it was possible for the management to be concentrated among a smaller number of people who resided in a particular area (typically, the county seat or a market town). Some of the joint stock associations secured a broad base of support. These were likely to be located in very rural counties. In Pike

County, for instance, the Board of Directors remained divided between the townships and 350 shares at $10 each were purchased by 150 people ("Pike County Report," 1871). Other counties sought the aid of a smaller group of wealthier men by issuing $25 shares. Those taking this approach, as did counties Henry and Vigo, got most of their support from inside the city ("Henry County Report," 1868; "Vigo County Report," 1868). Whether centered in large cities or not, agricultural societies sought to promote town growth far more than to encourage agriculture. Few changed their name to reflect the new priority.

Where leading citizens lacked investment capital, agricultural societies insisted that county governments ought to put the people's resources at their disposal. Why could not the Board of Commissioners purchase property for fairgrounds and then cede its management to the agricultural society? Using public funds, they maintained, was "the most feasible as well as the most equitable way of procuring" fairgrounds. An agricultural society, after all, was "organized for the benefit of the entire community. Why not then tax the real property of the community to buy that which is employed for their benefit and theirs alone" ("LaPorte County Report," 1868, pp. 293–294)? County officials may not have agreed that agricultural societies were properly categorized with schools as public institutions, but they had no difficulty with the line of reasoning. Using tax dollars to support a county fair was no different from using public money to subsidize a railroad line.

Indiana's most prominent fair had already set the precedent of using public funds to purchase fairgrounds. Hoping to attract the State Fair to Terre Haute in 1867, Vigo County had purchased 51 acres. Citizens' voluntary subscriptions to match the public donation yielded not one but two race tracks (1/2 mile and 1 mile) and first-class facilities ("Vigo County Report," 1867). To get the State Fair back, the Indianapolis City Council promised several thousand dollars and its railroad companies and citizens thousands more (State Board of Agriculture, 1869a, pp. 375–376). A great State ought to have a great capital city, and a great capital city had to host an exhibition. Did not it follow that a great county should have a flourishing county seat and a fair that did it justice? In Tippecanoe County (future home of Purdue University), two agricultural implement manufacturers joined hands with the Board of Commissioners to finance the purchase of 60 acres and facilities ("Tippecanoe County Report," 1871). Indiana's rural counties could not hope to match the scale of the donations offered by the state's largest cities—and their citizens could hardly hope to match the munificence of a John Purdue—but they could match them in public spirit. The General Assembly, controlled by Republicans, champions of energetic government action on behalf of economic growth, retroactively sanctioned their actions in 1873, granting counties permission to give agricultural societies a gift of $5,000 for making land purchases ("Act to Encourage," 1873).

With some counties combining public funds with private contributions to provide prospective investors a vivid testament of the spirit of enterprise in their communities, the rest presumably had little choice but to follow suit. In 1869 only 40 agricultural societies made reports to the State Board. The next year applications for recognition poured in "more rapidly than during any former period" (State Board of Agriculture, 1870, p. ix). Four years later, virtually every one of Indiana's 92 counties had a recognized county agricultural society, and an unknowable number of unaffiliated societies blanketed the state with fairs (State Board of Agriculture, 1874, p. 10).

The difficulties of traveling across Indiana's rural counties on its poor roads furnished townsmen with ample justification for duplicating the fair-hosting efforts. Besides, why should each county have only one thriving urban center, when two or three might be possible? Big fairs sprang up in little places that were tied to the world by scarcely more than a railroad line. Some 40 miles inland from the Ohio River, Jasper, the seat of Dubois County, had never boasted an agricultural society, although the county had "long felt the need." In 1871, the county got two, when citizens of nearby Huntingburgh decided to do their part ("Dubois County Report," 1871, p. 268). Landlocked, but located on the main route between Indianapolis and Lafayette, Boone County struggled for years to sustain a fair at Lebanon, its county seat; nevertheless, residents in the county's northwest corner thought they could do better, and did, at Thorntown ("Boone County Report," 1871).

Claiming geographic isolation, or negligence and mismanagement on the part of the county seat's fair-hosting efforts, was not necessary. Wayne County's agricultural society and fair at Richmond had always been at the forefront of agricultural quality and financial success. Its prosperity convinced residents of Cambridge City that the public demand for agricultural improvement was far from satisfied. Bounded by railroad lines on two sides, and only a five-minute walk from the city, the Cambridge City fair offered "every convenience" to exhibitors and visitors, and it claimed to draw its crowd from nine adjacent counties ("Cambridge City," 1871, p. 367). A few towns could claim, legitimately, that geographic barriers to travel made their fair-hosting necessary. Most were simply competing for a share in the fair-going market. The "unavoidable strife" of having several fairs within traveling distance divided "the interest and support of the public" and was "detrimental to the cause." Nevertheless, with the prospects of a town depending on a successful fair, the strife was "unavoidable" ("Greene County Report," 1878, p. 186).

The fair-going market was tight and the competition intense. Town boosters, therefore, put their hopes and dollars into their facilities, the one thing that might distinguish a fair from its rivals. The Johnson County fair at Franklin was revived in 1868 with a $10,500 investment in 30 acres, a racetrack, and a full complement of buildings ("Johnson County Report," 1869). Its backers had good reason for thinking big. Nestled in the corner of four rural counties (Johnson, Shelby, Bartholomew and Brown), a union fair held at Edinburg was drawing the crowds and the attention. Not to be outdone, the Edinburg Union Agricultural Society plowed its profits into new fairgrounds and produced facilities that matched Johnson County's in every respect save one. A 30-acre fairground might do well enough for the county seat at Franklin, but not for Edinburg. The union society purchased 80 acres ("Edinburg Union Agricultural Society," 1871). Neither place could claim to be more than a village. Their fair-hosting facilities rivaled the facilities found at the State Fair. The one-upmanship between Johnson County and Edinburgh Union is but one example of the scenario being played out across Indiana. If the funds could be raised, they were invested into expanded operations.

Since agricultural societies were paying close attention to what their counterparts were doing, a standard package for a town-boosting fair took shape quickly. Three or four one-story buildings of magnificent proportion housed the Livestock, Agriculture, Mechanical, and Miscellaneous display divisions. Floral Hall, ideally, was a distinctive two-story structure of octagonal shape, so as to accommodate a fountain or a towering dome decked-out with flowers. In the middle of the fairgrounds was sure to be a covered amphitheater, complete with raised seating, promenades, and columns, and large enough to accommodate several thousand people. Directly in front of the amphitheater was the indispensable race track, with a large bandstand (for public orators as well as musical entertainers) and a Dining Hall nearby. A few barns for livestock—with at least 200 stalls for horses and cattle, and half as many for sheep and hogs—were absolutely necessary, as were a handful of wells for watering. The show rings for displaying livestock, typically, were laid out across from the amphitheater on the far side of the race track. At least one small building was needed for the fair's officers and competition judges to carry out their business. Driving paths and walkways snaked between the permanent structures, leaving ample space for the temporary villages of tents and stands (for sideshows, food, and vending of various kinds) that sprouted during fair week. Around the grounds a stout fence, seven or eight feet high, complete with entrances and gates, ensured that everyone paid to witness the sights.

At the typical fair of the 1870s, a visitor might see virtually anything known to the era. The standard departments—Livestock, Agricultural, Mechanical, and Floral (or Domestic)—contained all that one would expect. Livestock typically included horses for general farm purposes, carriage, and saddle; cattle for dairy, beef, and fieldwork; sheep for different types of wool and mutton; swine of all kinds; poultry; and, the occasional animal of a more exotic species. Mechanical Hall was likely to be stuffed with farm implements and wagons; mills, pumps, and tiling machines; cabinetry and furniture; and, edge tools, pails, and leather goods of all kinds. Agricultural Hall was bound to contain vegetable monstrosities—two-foot long beets, rutabagas one foot in diameter, and, of course, a few 70-pound pumpkins—along with fruits, grains, and potatoes. The Domestic department would have canned fruits and jellies; rag carpets and woolen blankets; knit socks and mittens; homemade linsey and machine-fulled cloth garments; and cheeses and butter. The arrangement of articles, however, was seldom quite so predictable. Sometimes Floral Hall encompassed everything "in the domestic line," in addition to oil paintings, lithographs, engravings, old Indian relics and other "geological and archaeological curiosities" ("Lake County Report," 1875, p. 276). The Miscellaneous department might contain almost anything (sets of false teeth and stuffed birds were favorites). Plowing matches and farm implement trials often rounded-out the displays and competitions.

The feature attraction, inevitably, was the horse race. But, there was not just one kind, nor just one race per day. An endless number of heats took place on the track. Ladies rode in saddle and in buggies; they trotted, they paced, pulled stunts, collided, and raced. The gents did the same, on geldings and mares, in green runs and sweepstakes, in open categories and heats closed to men from elsewhere. With purses of $25, $75, $250 or more on the line, the tracks were "perfectly level and formed on geometrical principles" and meticulously scraped. The crowds gathered in immense throngs to see whose horse had enough "get up and go" to break 2:40 in the trot (Observer, 1870). With the close of the racing "died away the interest in the fair, by the spectators, and soon the grounds were almost deserted" (State Board of Agriculture, p. 141). Eager to please the crowds, agricultural societies extended the fairs throughout the week and increased the number of races held daily.

Whether on 30 acres or 80, agricultural societies set out to host fairs that matched their home pride and ambitions for the future. A big fair could not be had on the cheap. If an agricultural society already owned facilities, upgrades might cost a few thousand dollars. For those who started from scratch, the bill might easily exceed $10,000. No one expected to lose what he had invested. Quite the opposite, with the help of horse races and sideshows, everyone expected the fair's proceeds to cover start-up costs and yield a quick return. Some fairs managed to achieve this, but, in general, few could match the Loogootee Union District in seeing receipts exceed expenses by $4500 at its first fair ("Loogootee Union," 1872). Most shared the fate of the Delaware County Agricultural Society, to varying degrees. After investing $11,000 in fairgrounds at Muncie, its backers were rewarded with $7,000 of debt ("Delaware County Report," 1869).

Town-boosting ambitions ran amok everywhere, but nothing short of delusion can account for the creation of the Indianapolis Agricultural, Mechanical and Horticultural Association in 1871. The City Council appropriated $5,000; a leading railroad manager matched the public contribution with his personal funds; his company pledged another $10,000. With their backing, the association purchased 86 acres and erected facilities at the cost of $60,000. Their fair was a resounding failure. The gift of prescience was not needed to foresee that outcome for an event that began exactly one week before the State Fair. Such was the judgment of Indiana's best men in those heady days. One might suppose that Indianapolis's leading men would have learned something from the experience. If they did, it was the wrong lesson ("Indianapolis Agricultural, Mechanical," 1871).

Next year the Board of Trade joined hands with the Indianapolis City Council and the State Board of Agriculture to plan a 30-day State Fair and Industrial Exposition for 1873. Determined to keep pace with their rivals in Cincinnati, Louisville, St. Louis, and Chicago, they persuaded more than 400 of Indianapolis's leading businessmen to pledge $100,000 to a guarantee fund for the venture (Sutherland, 1873). Inevitably, the cost of improvements outran projections, but who could have predicted that Wall Street would get the jitters during the Exposition's first week? The financial panic of 1873 left the State Board of Agriculture holding the bag for its failed fair. Loaded-down by $40,000 of debt, it had no recourse but to continue hosting big fairs at Indianapolis in the hope of breaking even eventually. For the next 15 years, members of the State Board of Agriculture performed a biannual ritual of petitioning the General Assembly for an appropriation to meet the interest payments on its mortgaged fairgrounds (State Board of Agriculture, 1889, p. 78).

The State Fair's experience was replicated across the counties of Indiana. Already overextended, agricultural societies' ambition for growth turned to desperation. Seeking aid from nearby cities, some county societies moved their fairs into town and got bigger. Prior to 1874, the Allen County Agricultural Society had been more or less satisfied with a modest agricultural fair that, its secretary was proud to point out, "ignored fast horse rings entirely." That spring it reorganized, moved inside the Fort Wayne city limits, and built "a splendid race track" and facilities on 60 acres ("Allen County Report," 1870, p. 254, 1874, p. 60). So, too, the Elkhart County Agricultural Society decided its grounds two miles outside of Goshen were too remote and too small; with the help of $4,000 worth of subscriptions, it moved to a new site on the edge of town ("Elkhart County Report," 1874). A year deeper into the economic depression, the Montgomery County Agricultural Society thought it had a "reasonable prospect of success" of convincing the Board of Commissioners to purchase 60 acres at the county seat for its use. Retrenchment was not an acceptable option ("Montgomery County Report," 1875, p. 284).

Hoping to put their towns and counties on the map, agricultural societies had enlarged their facilities, premium awards, and prize purses. To sustain these in the midst of an economic depression, they needed to find ways to attract larger crowds. Fair managers pulled out all the stops, piling onto the program balloon ascensions, baby shows, velocipede races, foot races, and archery contests. Minstrel shows appeared alongside community brass bands. Town businesses put up money for "special" premium competitions to convince people to come to the fair. The ladies might turn out with jelly cakes or needlework for the opportunity to win $5 worth of goods from a merchant's shop. Who could resist the chance to find out who would be judged "the ugliest man" in the county ("Special Premiums," 1875)? The "citizens' special premium on fast-trotting," or on ladies' equestrianism, was sure to score big with the young men. Nothing boosted attendance like a horse race, so a full slate of "fast trotters, pacers, rackers and runners" was essential to "bring the crowds and keep them all the week" ("Jackson County Report," 1876, p. 141). As agricultural societies vied with each other to offer the largest purses to the fastest horses, enforcement of prohibitory regulations against gambling and the sale of intoxicating beverages relaxed. Wheels of fortune, crap games, "and other gambling contrivances were operated in broad day-light and within the observation of everybody" ("The Fair," 1875).

The commanding object was to draw crowds. To achieve it, fair managers set prudence and sensibility aside. In 1875, the Bartholomew County agricultural society booked Jefferson Davis, "just as Barnum would have engaged a giant, a fat woman, a six-legged calf, or any other monstrosity, solely as an attraction." The publicity stunt so enraged the community that they were forced to cancel the fair; their agricultural society collapsed and was forced to sell its fairgrounds to pay the mortgage (*History of Bartholomew County,* 1888, p. 37). The case of Bartholomew County is extreme, but countless others shared the same fate for more or less the same reason. Respectable opinion—that is, the town's moral crowd and the countryside's substantial farmers—had hardened against the horse racing and depravity sponsored at fairs. The vast majority of the population in most counties, farmers cast the deciding vote by refusing to attend degenerate fairs. In attendance and displays, the presence of agriculture at the county fairs dropped to its nadir. (The farmers' alternatives to the county fair are discussed in a later chapter.)

An agricultural fair with few farmers is a strange enough thing to contemplate, but an agricultural fair without livestock could not be permitted. The competitions of the county fair had once been closed to nonresidents in the hope of encouraging local farmers to improve. At the request of gentlemen farmers and local merchants in the late 1850s, the competitions had been thrown open to the world. After the Civil War, at the town-boosting fairs, all pretense of stimulating local farmers' emulative spirit was abandoned. Agricultural societies rolled out the inducements to persuade "the owners of the magnificent herds of blooded cattle to come from beyond the Ohio, and from the borders of the Mississippi, to break a friendly lance in a cattle tournament" at their fairs ("Tippecanoe County Report," 1876, p. 170).

In quick order, "a race of professional exhibitors [was] nursed into existence who make a business of traveling and taking premiums." With great cost and care, they assembled choice thoroughbred livestock and "pampered" them into a condition that defied "the competition of all ordinary breeders everywhere" (Hammond, 1872, p. 28). Typically, a handful of men swept the prizes for thoroughbred cattle, sheep, hogs, and other livestock. The Rush County Fair saw 16 coops of poultry (including chickens, ducks, geese, and turkeys) owned by one man take most of the ribbons ("The Fairs," 1877). A single Short-Horn cow, traveling from fair to fair, won 18 prizes in a single season ("Sallie Cleveland," 1879, p. 1). The fancy livestock on display at some fairs were so few that second place prizes could not be awarded "for the single reason of having no competitors" ("Dubois County Report," 1872, p. 310). Imported from the eastern states, Canada, and Europe, the purebred animals were the greatest in the land—just like the fast horses and the town boosters' ambition—but they offered little testimony to the quality of local agriculture. Only a few of the wealthiest farmers in each of Indiana's counties brought livestock to the fairs. Everyone knew it.

With fairs proliferating in the 1870s, professional exhibitors became stretched thin. There were too many fairs and not enough wealthy men traveling from place to place with purebred animals. Given the time and distance involved in travel, as well as the narrow window of opportunity for hosting fairs each autumn (between the harvest and cold, rainy weather), professional exhibitors could attend only a limited number of fairs. Unable to attend all fairs, they selected the largest fairs that could be fit into a schedule. The presence of fancy livestock at smaller fairs waned. Forced to compete with other counties to attract fancy cattle to their fairs, agricultural societies began offering greater inducements: the abolition of entry fees, more lucrative premium awards, even free grain and hay to spare professional exhibitors the trouble of "hunting up feed" for their livestock ("Lagrange County Report," 1874, p. 191).

Despite the growth of gate receipts from horse racing, the agricultural societies were offering more inducements for fancy livestock than they could afford. Faced with the choice of giving agriculture pride of place or attracting large crowds with horse races, agricultural societies made the sensible business decision. They gave more encouragement to the speed ring and less to the show ring. Premium awards for cattle (and other animals) were reduced and, often, entry fees were charged. Taxing cattle to finance fast horses was too much. With that decision, fair managers pushed the most influential farmers of Indiana's rural communities into the camp of fair reform (Thrasher, 1877).

Farmers had long complained about the fairs, on moral and practical grounds. Their complaints were easy for agricultural societies to ignore. Grumbling to a neighbor, or in a letter to the agricultural press, had no impact upon the design of the fair program. Farmers' absence from fairs diminished revenues, to be sure, but townsmen were confident that they could sustain the fairs without farmers' participation. And so they did, for a few seasons, with help from fast horses, young men, and town-dwelling crowds. Fair-hosting townsmen could not do without the aid of their counties' wealthiest farmers, however. As competitors in the livestock shows, these men lacked the crowd-pleasing power of the fast horsemen, but they made up for it with agricultural bona fides. They were not common farmers, but they were farmers nonetheless, and they commanded the respect of their neighbors. Now that the local livestock men had practical cause for complaint (in addition to moral discomfort from the fair's debauchery), they gave other farmers' grumbling coherence and direction.

There was a hard limit to the number of people a slew of horse races could bring out to the fair. By 1875, agricultural societies were discovering it. Complaining that their "financial embarrassment" was retarding their counties' economic development, agricultural societies petitioned the General Assembly to permit the county Board of Commissioners to purchase shares in their operations (State Board of Agriculture, 1875b, p. 24). The bailout request lacked contrition or apology for past errors. In light of the farmers' complaints about the fairs' operations, did townsmen really expect a legislature controlled by farmers (and Democrats, champions of limited government) to grant them an outright gift of tax dollars? In the halls of power, even more than on the fairgrounds, townsmen needed support from their counties' leading farmers. If agricultural societies refused to heed their demands for reform, the county fairs would fail entirely.

As Indiana's farmers withdrew their patronage and began to mobilize against the fairs' degeneracy, town boosters began to rediscover just how dependent their towns remained on the countryside. As late as 1880, farm families made up more than 60 percent of Indiana's population, while city dwellers comprised less than 20 percent. Employing less than 5 percent of the workforce (and providing only one-tenth the economic value of agriculture), manufacturing's share in Hoosier livelihood did not change appreciably during the 1870s (Thornbrough, 1965). Most of Indiana's counties were overwhelmingly agricultural; they could be promoted credibly on no other basis. A fair without much agriculture on display reflected poorly upon its county. The odds of success in advancing economic growth without at least some local farmers' support were long.

Fair-hosting townsmen were slow to come to grips with this inescapable fact about Indiana's development prospects. They had set their sights on attracting investment capital to jump-start manufacturing industries. Few farmers, however, subscribed to Governor Joseph Wright's doctrine of bringing the loom and the anvil into proximity with the plow. The economic turmoil following the Civil War had done little to change their minds. With railroad lines supplying outlets to market their wheat and hogs, farmers neither needed nor wanted enlarged town-dwelling populations situated nearby. With railroads supplying (relatively cheap) manufactured goods from abroad, farmers had little reason to support the growth of local industries. Boosting the town economy was a civic project, and farmers (on the whole) had not learned yet to be civic-minded. They did not perceive their own fates as being bound up with their nearby towns' economic prospects. Before Indiana's townsmen would get the help they needed to host successful county fairs, a critical portion of farmers would have to be convinced that they, too, needed the towns' economies to grow.

Reflections on Civic Learning

The pursuit of economic development became one of the defining features of the post-Civil War years in the Midwest. Some of Indiana's politicians invoked state pride (and western regional solidarity) against the moneyed interests of the East and on behalf of economic growth. Both themes, in their fashion, are forms of civism, but, for Indiana's townsmen, the claims of local community pressed more strongly. Townsmen did not whip up a campaign against Wall Street financiers or the federal policies that placed Hoosiers in thrall to the East. Nor did townsmen combine forces in state politics to compel the creation of agencies and policies that could have promoted abroad energetically the state and its economic prospects. Indiana's civic leaders pitted their economic ambitions against each other in a bidding war to attract investment capital and manufacturing facilities.

Whether state or local in emphasis, the overarching aim was the same. Conceivably, if local interests and rivalries had been subordinated to a statewide conception of the public good, the outcome might have been somewhat different. It is fantasy—not history—to suppose that Indiana's townsmen might have been persuaded to give a state agency responsibility for locating manufacturing projects in particular communities. Yet, channeling promotional efforts through a state agency that actively courted prospective capitalists and immigrants was a real possibility. The State Board of Agriculture's *Annual Report* was a poor substitute for a Bureau of Immigration. The Geological Survey was a poor substitute for public investment in internal improvements. Fair-hosting was a poor substitute for public policies that would have allowed incorporated towns to levy special taxes to attract manufacturing or to subsidize fledgling industries. Other states did these kinds of things; Indiana did not.

Part of the reason for Indiana's failure to follow suit is that the state government was burdened heavily with debts from the Mammoth improvement projects and the Civil War. Part of the reason is that civic leaders were unwilling to support measures that might have given other towns (particularly Indianapolis) more advantages in their pursuit of economic growth. Most of the explanation, however, resides in the simple fact that Indiana's farmers held the balance of power in the state's policymaking.

Other than railroad building, none of the pro-growth policies that were adopted in other states could gain the support of Indiana's farmers. Checked by farmers' representatives in the General Assembly townsmen sought to use agricultural societies and fairs to boost their communities' economic growth. In the late 1850s, county fairs had promoted growth by drawing farmers from the surrounding countryside into town for a few

days of extraordinary spending. After the Civil War, as rivalries intensified between towns, displaying a locality's advantages for relocation and
investment became the leading mission. Justification for the change was
grounded squarely in a rationale about the public good furthered by hosting showcase fairs. In its essential ideas about the role of manufacturing in
Indiana's development prospects, the argument was sound.

Making a solid argument about what will serve best the public interest
and obtaining popular approval is not the same thing. Most farmers did
not care whether towns had manufacturing or not and they did not want
their counties promoted as ideal sites for immigration. Elderly gentlemen
farmers and respectable younger farmers despised the moral degeneracy
that was sanctioned at fairs in the name of economic development. Their
refusal to attend and participate undercut townsmen's ambitions. Far more
than demonstrating a broadly shared commitment to economic development, the town-boosting fairs displayed just how wide the gulf had become
between Indiana's farmers and townsmen. Economic development was supposed to draw farmers and townsmen together into a harmony of interests.
Instead, it was pushing them apart.

References

Act to amend an act. (1861). *Indiana Laws* (41st Session), 2–3.

Act to encourage agriculture and agricultural fairs. (1873). *Indiana Laws* (48th
Session), 118–119.

Allen County report. (1870). In *Annual Report of the Indiana State Board of Agriculture.* Indianapolis, IN: State Board of Agriculture.

Allen County report. (1874). In *Annual Report of the Indiana State Board of Agriculture.* Indianapolis, IN: State Board of Agriculture.

Barnhart, J. D., & Carmony, D. F. (1954). *Indiana: From frontier to industrial commonwealth* (Vol. 2). New York: Lewis Historical Publishing Company.

A bill to enable counties. (1872). *Laws of the State of Indiana* (Special Session),
49–54.

Bland, T. A. (1867). Agriculture and manufacturing. *North Western Farmer, 2*(8–
9), 156, 174.

Boone County report. (1871). In *Annual Report of the Indiana State Board of Agriculture.* Indianapolis, IN: State Board of Agriculture.

Cambridge City district agricultural society report. (1871). In *Annual Report
of the Indiana State Board of Agriculture.* Indianapolis, IN: State Board of
Agriculture.

County fairs. (1874). *Indiana Farmer, 9*(34), 4.

Delaware County report. (1869). In *Annual Report of the Indiana State Board of
Agriculture.* Indianapolis, IN: State Board of Agriculture.

District fairs. (1866). *North Western Farmer, 1*(9), 133.

Donald, D. H., Baker, J. H., & Holt, M. F. (2001). *The Civil War and Reconstruction*. New York: W. W. Norton and Company.

Dubois County report. (1871). In *Annual Report of the Indiana State Board of Agriculture*. Indianapolis, IN: State Board of Agriculture.

Dubois County report. (1872). In *Annual Report of the Indiana State Board of Agriculture*. Indianapolis, IN: State Board of Agriculture.

Edinburg union agricultural society. (1871). In *Annual Report of the Indiana State Board of Agriculture*. Indianapolis, IN: State Board of Agriculture.

Elkhart County report. (1874). In *Annual Report of the Indiana State Board of Agriculture*. Indianapolis, IN: State Board of Agriculture.

The fair. (1875, September 11). *Washington Democrat.*

The fairs: Rush County fair. (1877). *Indiana Farmer, 12*(38), 4.

Greene County report. (1878). In *Annual Report of the Indiana State Board of Agriculture*. Indianapolis, IN: State Board of Agriculture.

Hammond, P. D. (1872). Report of the Tippecanoe Co. agricultural association for the year 1872. *North Western Farmer, 8*(1), 28.

Henry County report. (1868). In *Annual Report of the Indiana State Board of Agriculture*. Indianapolis, IN: State Board of Agriculture.

History of Bartholomew County, Indiana. (1888/1976). Columbus, IN: Avery Press Inc.

Indianapolis agricultural, mechanical, and horticultural association report. (1871). In *Annual Report of the Indiana State Board of Agriculture*. Indianapolis, IN: State Board of Agriculture.

Jackson County report. (1876). In *Annual Report of the Indiana State Board of Agriculture*. Indianapolis, IN: State Board of Agriculture.

Johnson County report. (1869). In *Annual Report of the Indiana State Board of Agriculture*. Indianapolis, IN: State Board of Agriculture.

A joint resolution in relation to publication of reports of the State Board of Agriculture. (1861). *Laws of the State of Indiana* (41st Session), 183.

Lagrange County report. (1874). In *Annual Report of the Indiana State Board of Agriculture*. Indianapolis, IN: State Board of Agriculture.

Lake County report. (1875). In *Annual Report of the Indiana State Board of Agriculture*. Indianapolis, IN: State Board of Agriculture.

LaPorte County report. (1868). In *Annual Report of the Indiana State Board of Agriculture*. Indianapolis, IN: State Board of Agriculture.

Loogootee union district report. (1872). In *Annual Report of the Indiana State Board of Agriculture*. Indianapolis, IN: State Board of Agriculture.

Marshall County report. (1867). In *Annual Report of the Indiana State Board of Agriculture*. Indianapolis, IN: State Board of Agriculture.

Montgomery County report. (1875). In *Annual Report of the Indiana State Board of Agriculture*. Indianapolis, IN: State Board of Agriculture.

Observer, The Thorntown fair. (1870, September 20). *Lebanon Patriot.*

Pike County report. (1871). In *Annual Report of the Indiana State Board of Agriculture*. Indianapolis, IN: State Board of Agriculture.

Richardson, H. C. (1997). *The greatest nation on the earth: Republican economic policies during the Civil War.* Cambridge, MA: Harvard University Press.

Sallie Cleveland. (1879). *Indiana Farmer, 14*(37), 1.

Special premiums. (1875, September 3). *Greencastle Banner.*

State Board of Agriculture. (1867a). Preface. In *Annual Report of the Indiana State Board of Agriculture.* Indianapolis, IN: Author.

State Board of Agriculture. (1867b). Proceedings. In *Annual Report of the Indiana State Board of Agriculture.* Indianapolis, IN: Author.

State Board of Agriculture. (1868). Proceedings. In *Annual Report of the Indiana State Board of Agriculture.* Indianapolis, IN: Author.

State Board of Agriculture. (1869a). Indianapolis fair between 1852 and 1869. In *Annual Report of the Indiana State Board of Agriculture.* Indianapolis, IN: Author.

State Board of Agriculture. (1869b). *Annual Report of the Indiana State Board of Agriculture.* Indianapolis, IN: Author.

State Board of Agriculture. (1870). Preface. In *Annual Report of the Indiana State Board of Agriculture.* Indianapolis, IN: Author.

State Board of Agriculture. (1872). Indiana state fair, 1872. In *Annual Report of the Indiana State Board of Agriculture.* Indianapolis, IN: Author.

State Board of Agriculture. (1873). Proceedings. In *Annual Report of the Indiana State Board of Agriculture.* Indianapolis, IN: Author.

State Board of Agriculture. (1874). Preface. In *Annual Report of the Indiana State Board of Agriculture.* Indianapolis, IN: Author.

State Board of Agriculture. (1875a). Introductory. In *Annual Report of the Indiana State Board of Agriculture.* Indianapolis, IN: Author.

State Board of Agriculture. (1875b). Proceedings. In *Annual Report of the Indiana State Board of Agriculture.* Indianapolis, IN: Author.

State Board of Agriculture. (1889). Secretary's report. In *Annual Report of the Indiana State Board of Agriculture.* Indianapolis, IN: Author.

Sutherland, J. (1873). President's address. In *Annual Report of the Indiana State Board of Agriculture.* Indianapolis, IN: State Board of Agriculture.

Thornbrough, E. L. (1965). *Indiana in the Civil War era.* Indianapolis, IN: Indiana Historical Bureau and Indiana History Society.

Thrasher, W. W. (1877). Do agricultural societies give cattle their proportional parts of premiums with horses? In *Annual Report of the Indiana State Board of Agriculture.* Indianapolis, IN: State Board of Agriculture.

Tippecanoe County report. (1871). In *Annual Report of the Indiana State Board of Agriculture.* Indianapolis, IN: State Board of Agriculture.

Tippecanoe County report. (1876). In *Annual Report of the Indiana State Board of Agriculture.* Indianapolis, IN: State Board of Agriculture.

Vigo County report. (1867). In *Annual Report of the Indiana State Board of Agriculture.* Indianapolis, IN: State Board of Agriculture.

Vigo County report. (1868). In *Annual Report of the Indiana State Board of Agriculture.* Indianapolis, IN: State Board of Agriculture.

5

Promoting the Farmer's Interest

Politics and the Grange

The Farmers' Movement and the Patrons of Husbandry

Throughout the 1850s, agricultural improvers called on farmers to join agricultural associations. Most farmers were apathetic. The 1870s witnessed the opposite. In a sudden burst of enthusiasm, farmers started forming clubs in their townships. Making two blades of grass grow where only one had grown before was not on their minds, but protecting themselves against town-dwellers' exploitation was.

From the Civil War's battlefields Midwestern farmers returned to their cornfields expecting prosperity. Hardship greeted them instead. Staple crop profits were hard to find. Too many farmers put the plow to the plains and too many railroads pulled the regions of East and West into a common market. Farmers knew that there was more involved, though, than a mismatch between agricultural supply and consumer demand. Through organizations, townsmen were taking advantage of the farmer's isolated condition as an independent producer and consumer. Railroad companies conspired to protect themselves from competition and to charge exorbitant freight rates. Manufacturers formed rings to corner the market and fix the

Civic Learning through Agricultural Improvement, pages 97–121
Copyright © 2011 by Information Age Publishing
97

prices of farm implements and other goods. Merchants, agents, and warehouse operators—the middlemen who stood on either side of the farmer's contact with the commercial economy—did the same. Town-dwellers' success at "co-operative effort," one aggrieved farmer thought, "should teach the farmer the lesson" (Chase, 1873, p. 159).

It did. Declaring themselves "tired of calm submission to all kinds of extortions and indignities," farmers started organizing (Ratliff, 1873, p. 128). They sent letters to the agricultural press describing how they were banding together. Requests for information poured in from other farmers who were seeking tips on organizing strategies and ideas about what ought to be done (Doty, 1873, p. 109). Schoolhouse meetings grew into mass conventions and picnic rallies attended by thousands. Singing songs that celebrated agriculture and called for justice, farmers waved flags and paraded with banners. Gathering in groves, fairgrounds, and halls they heard speakers decry monopolies, middlemen, and corrupt politicians (Harvey, 1873; "Mass Meeting," 1873). July 4, 1873 was the Farmer's Fourth in the Midwest. It was just the beginning. Through the next year, the discontent mounted as farmers and their office-seeking champions brought their grievances to bear on the political establishment in the election campaign of 1874.

What was the farmers' movement all about? At its heart were low farm profits and an old civic goal: bringing the producer and consumer closer together. But, it was not the home market (county seat) that captured farmers' attention. When farmers spoke about bringing the producer and consumer closer together, they referred to the span between the rural West and urban East. As producers of grain and meat, farmers wanted railroad freight rates lowered and steps in handling eliminated so that they could earn higher profits. As consumers, farmers wanted the implements and household goods that were obtained most cheaply by importing them from distant manufacturers. As both producers and consumers, farmers wanted public policy action that would reduce the railroad rates and shipping costs that impeded commerce between West and East.

Farmers' pursuit of these aims conflicted with the economic interests and civic aims of nearby town-dwellers. The more direct and distant the farmers' commercial transactions, the less farmers utilized the merchants and agents residing in their county seats. The more farmers purchased from distant suppliers, the less they supported local manufacturing. Farmers' savings deprived nearby towns of their economic growth potential and (some) town-dwellers of their livelihood. Coupled with farmers' demands for governmental regulation of the railroads—which were the towns' lifelines—their goals were antagonistic to the civic ambition to grow civilization in the Midwest.

The terms, grangerism and the granger movement, are often used to denote the farmers' campaign against the abuses of merchants, middlemen, and railroad monopolies in the early 1870s. As intended by its founders, the Grange (officially known as the Patrons of Husbandry) was supposed to be an association for educational, social, and fraternal purposes. Grange-joining farmers, however, expected to do more than to study tillage and socialize; they brought economic cooperation and political mobilization into the organization as means of advancing their interests as farmers (Barns, 1967; Buck, 1921; Woods, 1991). Almost instantly, the Grange became the largest farmers' organization in the United States. But, the Grange was not synonymous with the farmers' movement; in design, at least, it was the equivalent of a civic organization for farmers. And as one might expect from a civic organization, according to Grange teachings, when understood and acted upon rightly, the economic interests of farmers and townsmen were mutually beneficial, not antagonistic. The first order of business for Grange leaders, therefore (at least for those who sought to be respectable citizens and honorable grangers) was to prevent their organization from being used as a political weapon—as an economic interest-based farmer's party—in a destructive crusade against non-farming enterprises.

Urging Farmers to Advance Their Interest

The ground was well prepared for the Patrons of Husbandry. In the late 1850s, dissatisfied with the county agricultural societies, improvement-minded farmers began creating clubs in their townships. Some were clubs only in the sense that individuals subscribed to a common agricultural paper. Others met periodically to discuss practical agriculture: manuring, subsoil plowing, drainage, and livestock care were likely topics. Visiting each other's farms and holding strictly agricultural fairs—that is, displays of farm products without horse racing and sideshows—rounded out their activities. How many of these farmer clubs existed is unknown, but they were dispersed across the countryside on the eve of the Civil War (Demaree, 1941; Mills, 1871).

To preserve harmony in their meetings, most township clubs declared political discussion off-limits. They presumed that every farmer, if he studied the science and economy of agriculture, would perceive rightly his true self-interest. The farmer's true self-interest was not confined to the farm: marketing conditions and transportation costs extended it outward into society. On this basis, champions of agricultural improvement claimed that their favorite policy measures, such as railroad building, manufactures, and road improvement, were not political issues, but agricultural matters. Other farmers' failure to recognize their true self-interest was revealed in their apathy toward public policy related to economic development. Its main result was unfavorable economic conditions for the inhabitants of town and country alike.

Facing extortion from railroads and middlemen after the Civil War, farmers got the message to pay attention to public policy. Disgruntled farmers had little difficulty turning a scheduled discussion on subsoil plowing or crops into a discussion that ended with the resolution: "All we ask is equal rights and justice" (Wennick, 1873, p. 159). Had not farmers been told for decades that agriculture was the foundation of prosperity? The logical counterpart was that unless agriculture prospered other industries could not fare well. Had not farmers been told that all sectors of the economy were mutually dependent for advancement? If so, all producers—regardless of their occupational line—were supposed to receive a fair rate of return on their labor and investment. This was not occurring. Men in mercantile, manufacturing, and shipping pursuits received far greater returns than even the most cost-conscious farmers ("Capital and Labor," 1874). Farmers' conclusion about the earnings disparity was obvious and largely accurate. Through combinations and illicit agreements townsmen were distorting the natural workings of the free market economy and taking advantage of farmers who, as isolated individuals, lacked bargaining leverage when purchasing and selling in commercial markets.

For mutual protection farmers needed their own organizational system. How else could they coordinate their purchasing and selling activities to achieve economy of scale comparable to that enjoyed by townsmen who were affiliated with corporate networks? How else could farmers channel their opinions into a common lobby that could bring agricultural interests to bear on public policy? The existing county agricultural societies could not serve these purposes, they were controlled by leading townsmen. The Patrons of Husbandry could. It had an institutional structure that reached from the neighborhood to the State to the Nation (but, oddly, omitted the county) and ready-made provisions for recruiting farmers. Aided by the agricultural press, traveling lecturer-organizers had little trouble attracting notice. The spring of 1869 saw 40 granges in Minnesota and a handful more elsewhere, and by the close of 1873, local granges were organized in all but four states. Two years later, three-quarters of a million people were dues-paying members (Nordin, 1974). Indiana alone had 65,000 members organized into more than 2,000 local granges. The Grange enjoyed, as one participant recalled a few years later, a growth rate "unprecedented in history and unequaled by any order before known to the world" (Indiana State Grange, 1875–1876, p. 26).

But, what, exactly, was the Patrons of Husbandry? Founded by "one fruit grower and six government clerks, equally distributed among the Post Office, Treasury, and Agricultural Department," the Grange was a strange concoction (Buck, 1921, pp. 3–4). It picked up the older agricultural societies' mission to promote diversified agriculture and improved business methods. To this agricultural mission, which always had the civic mission of drawing farmers closer to towns and the ethos of progress, were added the elements of a fraternal order. Ritual secrecy and symbolism were copied from Freemasonry and infused with Christian and agricultural aspects. The Grange contained much for farmers to appreciate: prayers, songs, harvest festivals, and various ways of celebrating the virtues of agriculture. It also contained much that farmers would find objectionable. Ornate regalia, elaborate ritual, and a hierarchy of ranks were not traditionally associated with agriculture. As the Grange spread, Midwestern farmers added a cooperative dimension for selling farm produce and purchasing consumer goods.

Impressions as to what the new farmers' association might do, flooded into the agricultural press. Some men thought the Grange was simply another agricultural society. Others thought it was a cooperative association for obtaining farm machinery at reduced prices. Perhaps the whole business was a scam? After all, if the Grange's founders "had been so terribly exercised" about monopolies and middlemen, why did they conjure up a "system of tintinsel ritualism" and "mystical pageantry to peddle to the farmers" ("From an Objector," 1873, p. 65)? Other men expected immediate and assertive action to make "the great world of non producers" more appreciative of the farmer (Deal, 1873, p. 42). They knew of no other genuine farmers' organization, so they were ready to "go into this one" even if, as rule, they were "opposed to secret political societies." The editor of the (Indianapolis-based) *North Western Farmer* was quick to admonish contributors that they "should not call the Patrons a political society" (A. C., 1873, p. 66).

The Grange was not intended to be a political organization. It followed the precedent established by older agricultural societies: agricultural matters (and grange ritual) inside the gate, politics outside. Its constitution stated explicitly that "political questions will not be tolerated as subjects of discussion in the work of the Order" (Indiana State Grange, 1873, p. 18). Separate spheres for farmer talk and citizen talk could not hold though. Too much of what concerned agriculture was not on the farm. Instead of banning political discussion, farmers divided their meetings into closed grange sessions and open sessions that dealt with public policy. A narrow interpretation of what qualified as political questions (candidates seeking office and issues upon which the political parties had taken stands) left ample leeway for discussing farmers' grievances against railroads, middlemen, merchants, bankers, and just about anything else related to the business side of farming. To translate talk into action, farmers grafted a new organizational feature onto the Grange: county councils. Copied from the more assertive Illinois State Farmers' Association, the county councils were not recognized officially as institutions of the Grange. They could—and did—advance farmers' interests as they saw fit, without fear of sanction. Taken as a package, these measures enabled Midwestern farmers to use the Grange for their education in political economy and public policy (Collett, 1874; "Progress of the Order," 1873).

The railroad monopoly question (and the middleman problem) was the prevailing subject of interest at farmers' gatherings. Since the late 1850s, local rail lines had been increasingly consolidated by eastern-owned companies. With each round of consolidation, freight rates went up. The railroad companies charged higher rates for short hauls between lesser cities than for long hauls between major urban centers. In effect, they used the natural monopolies possessed by single lines running through remote areas to subsidize low rates on major lines that experienced competition from rival corporations. Farmers resented this unequal treatment that deprived them of the little profit obtainable from staple crops. Most wanted (at least) the smaller rail lines restored to local ownership, which presumably would be more responsive to the farmers. Alternatively, state legislation might compel railroad companies to reduce rates and to set proportional schedules for mileage, bulk, and weight. Between 1867 and 1873 all legislative proposals to grant Indiana's General Assembly the authority to revoke railroad charters and to fix rates failed (Thornbrough, 1965).

As winter turned to spring in 1873, Illinois took decisive action. Under pressure from the State Farmers' Association, the Illinois General Assembly established a commission for setting rates and penalizing violators. Other states copied the new granger law. Indiana did not. A bill for "regulating and equalizing freights" passed the State Senate, but mysteriously got "buried up and lost sight of" in the House. Despite hopes that a little farmer-agitation might bring the bill "out from its hiding place," the bill failed to reappear. Instead, following Congress's lead, Indiana's General Assembly engaged in a "salary grab." In another move that was well calculated to spark farmers' interest in public affairs, it tripled the property tax rate and transferred assessment from township to county officials ("Farmers Should Organize," 1873, pp. 19–20). The Indiana General Assembly would not convene again for two years.

Denouncing the railroads' extortion and the General Assembly's indifference, Hoosier farmers vowed to "make all other questions subordinate to the railway monopoly question" (Conner, 1873, p. 87). Crop prices were at their lowest rate in living memory; freight rates and taxes were at their highest. Farmers did not need to be reminded that they had "none but themselves to blame" for failing to "take a proper interest in public affairs" ("Farmers Should Exert," 1873, p. 113). While fields were plowed and fences repaired, farmers organized. They flocked to hear speeches by the leaders of the Indiana State Grange, the Illinois State Farmers' Association, and the editors of the farm papers. Popular outrage at the salary-grabbing, tax-raising, and railroad-monopoly-favoring legislators mounted. Few doubted that the next legislative session would witness a marked change in lawmakers' behavior. If farmers united "in calling for any practical reform," their "combined opinion" would be an "irresistible" force ("Greencastle Banner," 1873, p. 5).

Grange leaders repeatedly urged farmers to "cut loose from all party ties and pledge themselves to vote only for those men and those measures that will tend to advance their interests" ("The Patrons and Politics," 1873b, p. 159) As a popular movement, grangers envisioned their organization as a purifying force in political affairs. Farmers represented the majority of the nation's population. Asserting the farmer's interest by seeing that elected officials acted upon agricultural concerns, therefore, served the public interest. Farmers who voted irrationally by party affiliation alone, rather than based upon their economic interests, left public officials free to act for partisan, factional, and personal advantage.

When they invoked farmers' nonpartisan economic interest, grangers followed precedent. Throughout the antebellum era, the most respected citizens of town and country had called on farmers to recognize their self-interest and to act accordingly. Their reasoning was utilitarian: whether encouraging railroad building or diversified agriculture, a policy related to agriculture was considered to be above party if it promised the greatest good to the greatest number. The 1840s and 1850s were ripe for a utilitarian ethos. Before railroad networks pulled far-flung places into an integrated economic zone, communities were relatively isolated. Few people considered the possibility that some places—and a small number of people—might benefit far more than others from economic development policies that were general in nature. Besides, farmers had proved (time and again) that they were reluctant to get involved in organizations. Who could conceive that farmers might act collectively on their self-interest in ways that ran contrary to economic growth?

After the Civil War, those conditions no longer existed. Extensive railroad building ended community isolation; currency contraction and public debt directed the flow of money eastward; the opening of the Great Plains ended agricultural prosperity; and, the explosive growth of the Grange proved that farmers could organize on a broad scale. Established precedent no longer applied. Of itself, even a program of moderate policies that were designed to favor farmers might jeopardize Midwestern towns' economic development. The prospect of a union of farmers advancing their economic interests through direct political action raised eyebrows. It also presented a tantalizing opportunity for men who did not mind inciting a little anti-town sentiment among the farmers as a way to gain public office.

Restraining Farmers from Advancing Their Interest

Financial panic struck in September of 1873. Packing houses and mills closed, railroad workers went on strike, crop prices sank lower still, and the flow of farmers into the Grange swelled. Farmers began to prove that they were quite capable of thinking about politics in terms of interest rather than party. Under various names (Anti-Monopoly, Independent, Farmers' and People's), political parties that claimed to represent the farmer emerged in Illinois, Iowa, Minnesota, and Wisconsin. Prominent grangers—their identity as such hardly secret—stood in the front ranks. As the popular press scooped rumors about agrarian and communist ambitions, muted concerns about the farmers' movement turned into alarm (Nordin, 1974).

For most grangers, including those who sought political office, urging farmers to act on their interest was in no way intended to be an invitation to radicalism or class warfare. Only legitimate means of reform, the petition and the ballot box, were advocated. The prospects of violence and destruction of property (the dangers most often feared in popular movements) were remote. Grange picnic rallies included women and children; they were orderly and sober, unlike the typical election campaign event. Prominent grangers' concern fell far short of the Eastern press's wild imaginings of an agrarian communist uprising. The real worry was that farmers would insist upon legitimate, if unwise, political actions that would be harmful to Indiana's towns.

Moderation had to prevail. Mobilizing farmers against property rights and economic development was one thing; seeking measured reforms to curb the economic abuses of (some) middlemen and (externally owned) railroads were another. Repeatedly, leading grangers warned that it was "not desirable to destroy or even to interfere with the real interests of either the merchants or the railroads." Merchants were "necessary for the transaction of business"; railroads, "when properly and honestly managed [were] capable of the greatest good to the country." Since they performed

essential services, there was no reason why mercantile and railroad operations should not yield "a reasonable per cent in return under careful management" (Billingsley, 1873, pp. 6–7). To concede this was not to betray the
farmer's interest, rather, it was to grant to others what farmers demanded
for their own. If they wanted townsmen to perceive "the justice of our demands," farmers needed to be "fair and just" in their strength (Stevenson,
1873, p. 2).

The calls for moderation were genuine. With the financial crisis growing more severe, farmers were susceptible of being easily led by well in-
tentioned (but wrong-headed) men. Worse still, the "belligerent talk" and
"wild unreasonable threatenings" voiced by farmers at rallies and country
stores invited the attention of demagogues (Templin, 1874, p. 2). In a highly
politicized and closely divided state like Indiana, men who lacked scruples
could use farmers' discontent to great advantage. The 1874 state elections
were a full calendar year away, but the season for nominating candidates for
public office was just around the corner. Traveling to midsummer farmer
rallies, the State Lecturer had already witnessed men trying to convince
farmers "to follow some one political idea" ("Mass Meeting," 1873, p. 5). By
autumn, office-seekers of all stripes were "running about trying to rent a lot
to sow to turnips so they might have the requisite qualification to become
'grangers'" (Templin, 1873, p. 3).

By December, politicking within the granges was widespread. A stern
warning from the State Grange's leadership about keeping partisan politics
out of the Grange had no effect. The *Indiana Farmer* editors (Patrons themselves) took the State Grange's officers to task for not doing more. Clearly,
"denial alone" was insufficient to convince the public that the new farmers'
organization did not intend "to play a considerable part in the future politics of the country" ("The Patrons and Politics," 1873a, p. 6). Farmers were
joining with the intention of "pressing it forward for political purposes"
and there was no shortage of political aspirants who were ready and willing "to do or die 'for the cause of the farmer.'" The editors of the *Indiana
Farmer* could only express their hope that Patrons would turn their backs
on the "unknown political adventurer without experience and of doubtful
character" and take the "higher ground" established by the Order ("Higher Ground," 1873, p. 8).

Exactly what that higher ground represented was unclear. Despite the
Order's official prohibition against bringing partisan politics into grange
meetings, leading grangers were in the thick of electioneering in their
home communities. A national convention in February of 1874 provided
the opportunity to clarify the Grange's relationship to politics. There, the

Patrons approved a Declaration of Purposes, a high profile statement designed (in addition to instructing farmers) to reassure anxious town-dwellers that the Grange was led by sensible, conservative farmers. The Patrons of Husbandry would "wage no aggressive warfare against any other interests" on farmers' behalf.

The Patrons declared "emphatically and sincerely" "that the Grange—National, State, or Subordinate—is not a political or party organization. No Grange, if true to its obligations, can discuss political or religious questions, nor call political conventions, nor nominate candidates, nor even discuss their merits in its meetings." With that, the Patrons disassociated itself from all political parties and any men who tried to use its meetings for partisan purposes. For good measure, the convention refurbished the Jeffersonian doctrine, "never seek or decline office," into the slogan: "THE OFFICE SHOULD SEEK THE MAN, AND NOT THE MAN THE OFFICE." Broadcast throughout the nation in farm papers and grange meetings, the slogan put the onus of suspicion on men who were seeking office for the first time. Whether they were truly friends of the farmer or not, they would be compelled to defend their purity of motive.

The Patrons of Husbandry was out of politics. It was not about to deny the farmer his right to participate though. As a citizen, it was the farmer's duty "to take a proper interest in the politics of his country." Informed by Grange principles (which "underlie all true politics"), the farmer would seek the election of "none but competent, faithful, and honest men, who will unflinchingly stand by our industrial interests." As a politically active citizen, he would "do all in his power legitimately to influence for good the action of any political party to which he belongs." If all farmers did these things, they could "purify the whole political atmosphere" and ensure that public policy reflected the common good. Their actions, however, would be taken as citizens, not as members of the Patrons of Husbandry (Indiana State Grange, 1889, pp. 53–56).

The Declaration of Purposes was about as clear and direct as a voluntary association's position statement on its role in popular politics can be. Yet, the Patrons of Husbandry could not discipline its members' political behavior. It could not prevent farmers from joining farmer parties or independent campaigns that—like the Grange—pledged to purify political affairs and to advance their interest. It could not prevent farmers (as farmers and as citizens) from voicing anti-town sentiments or advocating ill-conceived economic policies. As individual citizens (who happened to attend grange meetings), farmers could do these things; in the run up to the 1874 state elections, Midwestern farmers were.

In January of 1874, more than 267 subordinate granges were registered with the Indiana State Grange, with an average of three clubs per county. Throughout the winter and spring, the pace of grange organizing accelerated (Indiana State Grange, 1875–1876). There was no denying that it carried a backlash against the political establishment. Nor was there any means to prevent it, for the idea of following the Illinois State Farmers' Association in creating an independent (farmer) ticket was gaining ground. In April, two separate events tipped the balance. The first was that court house rings (that is, wealthy residents of the county seats) selected Democratic and Republican candidates for county offices without consulting the countryside's farmers ("Parties," 1874). The second was President Grant's veto of a bill to increase the paper currency supply.

Characteristically, when nominating candidates, the court house rings did not consult farmers; comparatively few farmers attended the conventions. This election year was not a normal year. Leading townsmen failed to take note. Cutting farmers out of the nominating process was a direct invitation for them to organize their own slate of candidates. This seems straight forward. By contrast, expanding the paper currency supply might seem to be an unlikely issue to spur an independent campaign. Traditionally, farmers preferred hard money. The $433 million in greenbacks issued during the Civil War softened their opposition to paper. As the greenbacks were retired in subsequent years, the Midwest grew short of currency, particularly at harvest. The shortage of money compelled farmers to accept low prices for their crops at harvest (because other farmers were looking to sell, too). Then, as the demand for circulating media slacked through the winter, prices rose on the goods farmers had to purchase. From hard experience, farmers began to perceive the value of artificially adjusting the currency supply to respond to seasonal fluctuations in the volume of commerce. Many farmers continued to prefer gold, but their opinion of greenbacks went up markedly.

The farmers' change of heart was shared by Midwestern bankers and manufacturers. They saw currency shortage as the key barrier to economic growth. After financial panic struck in 1873, Indiana's leading bankers and businessmen proposed an elastic currency. Essentially, their plan called on Congress to regulate directly the currency supply (as distinct from indirect manipulation through Treasury bonds and national bank notes). To expand the supply immediately, the plan called on Congress to authorize Treasury bond holders to exchange them for legal tender notes. On the whole, the proposal was a moderate move toward currency inflation. Touted as the "Indiana Plan," it spread among the populace as the notion that Congress should issue more greenbacks (Thornbrough, 1965, pp. 286–287). The Indiana Plan became one of the planks in the *Indiana Farmer's* answer to the question: "what does this movement of the farmers mean" ("What Does This," 1873, p. 8)?

The currency bill passed by Congress in the spring of 1874 proposed restoring the supply to a level near that of the Civil War years. It was widely approved by resolutions passed in granges and state legislatures throughout the Midwest. President Grant's veto, and Congress's failure to override it, set off a wave of condemnations, for favoring wealthy capitalists and eastern interests over the common farmer ("Congress," 1874). In Indiana, the call went out for the "friends of reform" to gather at Indianapolis to devise an independent platform and to nominate candidates who would advance the farmer's interest ("To the County and District Voters," 1874, p. 4).

The Independent Challenge in the 1874 Elections

The *Indiana Farmer* carried the announcement and explained the rationale for holding the Tenth of June convention. President Grant's veto and the county nominations had convinced "thousands of the best men of the State" that "the only way to purify the country" was through "an uprising of the people without regard to former parties." Simultaneously, editor and owner, J. G. Kingsbury, made it clear that Indiana's leading farm paper and farmers' organization had nothing to do with the independent convention. The *Indiana Farmer* opposed it, and the Patrons of Husbandry, "as an Order, were in no way responsible" for it ("The Tenth of June," 1874, p. 4).

Kingsbury opened the columns of the *Indiana Farmer* to those who were skeptical of the intentions and prospects of the independent movement. Would not the formation of a class party representing the farmer unite other interests in society against it? Was it even conceivable (in an age of hard core political affiliations and bloody shirt waving) that typical citizens would "lay [party] prejudice aside" to join together "and tell the old political hacks and wire pullers to stand aside?" Was it not more prudent to "purify" the established parties, by seeing that "none but men of honesty and capability occupy positions of trust and honor" (Mace, 1874, p. 1)? Even if an independent party should triumph at the polls, was it not folly to assume that men chosen to advance the farmers' interest would be more virtuous than previous officeholders (Granger, 1874)? Others doubted that the independent campaign was genuine; the whole thing, they thought, would be turned over to the Democratic Party. One officer of a subordinate grange even opined that the so called "farmers' call" was nothing but a "political move" by State Grange officers who, he claimed, hoped to secure "positions on the State ticket" (McCoy, 1874, p. 5).

Beyond the skepticism, the difficulty facing the Tenth of June convention was to find new men to take the place of the old in public office. Most men who were willing and able to serve were tainted by affiliation with the existing political parties. Were such men trustworthy? Of the men who had remained outside of partisan politics, those most likely to be capable officeholders were leading grangers. Some were willing to offer their names, but most were not. They opposed the movement, were unable to leave their farms to assume public duties, or did not want their character stained by the charge of opportunism. To fill the positions, the independent movement would have to rely upon non-farmers without established reputations. Their burden would be to prove that they were friends of the farmer without alienating the town-dwelling portion of the citizenry.

The Tenth of June Convention failed to meet expectations. About 500 men attended. In addition to some prominent grangers, a number of "experienced politicians of both political parties" attended with the intent of making it their mission "to effect a failure, in order to perpetuate old party domination." Only one of the candidates chosen to run for State Office was an actual farmer; a pork-packer, a carpenter, a publishing company agent, and two lawyers rounded out the state ticket. The nominee for Secretary of State had been expelled from the Odd Fellows for embezzlement; the nominee for Auditor was a sitting member of the Democratic Central Committee. The nominees for Treasurer, Attorney General, and Superintendent of Schools were virtually unknown. Only the nominee for Judge of the Supreme Court had a statewide reputation; he declined. Delegations from several districts refused to support the nominations ("Reform," 1874, p. 4).

The Independent platform fared little better; far from being a radical assertion of the farmer's interest against townsmen and railroad corporations, it lacked conviction. The preamble was a cut-and-paste job that placed the stridency of the Farmer's Declaration of Independence (of the Illinois State Farmers' Association) alongside conciliatory passages lifted from the Patrons of Husbandry's Declaration of Purposes. The platform offered little by way of substance. The Indiana Plan for currency inflation was adopted, naturally, but nothing was said about regulating railroads or revoking charters. Although temperance was a favored granger cause, the platform waffled on how the right of the people to regulate the liquor trade might be put into operation. Only the General Assembly's recent salary grab and property tax were condemned forthrightly. These were easy targets, of course, and no appeal to farmers could be complete without a pledge to reduce public expenses and taxes to the "lowest possible limit." Despite the disappointing results, the *Indiana Farmer* suggested that "the other party managers will be wise to heed" the Independent platform ("The Independent Convention," 1874; "Reform," 1874, p. 1).

The advice was unnecessary. The Republican Party convened a week later. It, too, appropriated phrasing from the Patrons' Declaration of Purposes and pledged to do more for the farmer. Beyond the usual Republican themes on behalf of diversified economic development (which were part of Grange doctrine), the platform pledged that "relations between capital and labor" would be adjusted properly, "in order that each may receive a just and equitable share of profits"; to that end, corporations would be held "in strict conformity to law." On the railroad monopoly issue, not a word appeared (a pledge to end land grants was included). On the currency question, Republican delegates settled for a vague pledge to adjust the supply "as may be necessary to meet the wants of the agricultural, industrial and commercial interests of [all sections of] the country." On temperance, the right of the people to implement local option regulations was honored. Naturally, to appeal to farmers, pledges were issued to reduce taxes, salaries, and the powers of county governments (Henry, 1902, pp. 47–49). Ultimately (as party managers well knew), nothing on the platform mattered all that much for the current election cycle. Discredited at the national level by corruption scandals and southern reconstruction, and discredited in Indiana by its leadership of the late General Assembly, the Republican Party's program was a lost cause.

On the eve of the Democratic Party's convention, the *Indianapolis Sentinel* (the state's leading Democratic paper), reported that a "large number of republican patrons [were] prepared to support the democracy if the ticket be at all good on account of their opposition to continuing the same men in office" ("The Word," 1874). Out of power for most of the past 15 years, the Democratic Party could heap blame on the Republicans. It could also nominate farmers for public office and respond to their grievances in ways that the Independent Party could not. Long-established, with broad support across town and country, the Democratic Party could not be accused of being a class party. The Republicans were demoralized, the Independents of dubious character. The Democrats had all the advantages.

Shrewdly, the Democratic Party's managers put off their convention until mid-July. The delay fueled rumors as to whether the Independent Party was genuinely an independent campaign. Grangers contributed to the confusion. Determined to vote by the man and not by party, they used their (unofficial) county councils to nominate independent tickets for county offices and legislative seats. In addition to candidates for the Independent Party, the county councils selected (from the two established parties) the candidates found to be "most unexceptionable" and permitted double-endorsement ("The Independent Ticket," 1874). The grangers' approach was, at best, a mixed blessing for the Independent Party. Ultimately, it favored the Democratic Party: farmers were not forced to abandon their traditional party, since candidates could run on both tickets. In the meantime, while the Independent Party's nominations for lesser offices were plagued with uncertainty, word of its disappointing platform circulated among the citizenry.

The delay also gave the Democratic Party time to devise an effective platform. Two resolutions dealt with railroads, five with the currency, and five with public finance and government. In contrast to the Independent and Republican silence on railroad monopoly, the Democratic stand was crystal clear: "railroads and all other corporations...should be rendered subservient to the people's good"; legislation was needed to protect the "producing interests of the country against all forms of corporate monopoly and extortion." On the currency question, Democrats joined Republicans in waffling: they advocated both "soft" and "hard" money by demanding an elastic currency supply (with greenbacks), and insisting on a return to specie payments on treasury bonds "as soon as the business interests of the country permits." To no one's surprise, the platform called for "severe retrenchment, reform, and economy in all branches of the public service." Repealing the Republican-passed salary grabs and property tax increases, as well as restoring the traditional township-based system for assessing property values, were essential priorities. Low taxes, limited government, and local authority were traditional Democratic themes (and part of Grange doctrine). Farmers would have no difficulty understanding those positions.

The Democrats left no stone unturned. They waffled on temperance. They played on race hatred, condemning efforts to secure equal rights for African Americans in the South and denouncing interracial schooling (in a state with few black citizens). They pandered to the farmers' movement by extending "sympathies and support" to efforts of "the industrial classes to ameliorate their condition" through "the formation of associations for mutual protection and advancement." With that gratuitous nod and a heartfelt pledge to honor pension obligations to Civil War veterans, widows, and orphans, the Democratic Party closed its platform. A sweeping victory was all but inevitable ("Launching the Craft," 1874, pp. 3, 8).

The Independents would get far less help than they had hoped for from the farmers' movement. Their party was unable to present itself as the farmer's party. Candidates of all persuasions were using local granges to mobilize farmers. By July, more than 1,900 granges (on average, roughly 20 per county) were registered in Indiana (Indiana State Grange, 1875–1876, p. 26). Grange-sponsored rallies were the main forums for getting the message to farmers. At those rallies, speakers were invited from the several political parties to rail against the "money capitalists, rings, and monopolies," to sing the virtues of hard money and retrenchment, or to urge farmers to vote for good honest men ("The Grangers in Motion," 1874). At least one prominent Granger was sure to remind farmers that the Patrons of Husbandry was neither a political organization nor affiliated with a particular political party. To reinforce the point, electoral politics was only one of the topics for public-speaking: agricultural economy, practical agriculture, and agricultural education rounded out the program ("Grange Pic-nic," 1874).

The summer rallies were politically driven, but, ultimately, they were farmer rallies, more so than partisan campaign events. Processions a mile or two in length kicked off the festivities. Agricultural enthusiasm was displayed with oxcarts bearing agricultural products, floral displays, and banners with harvest scenes etched upon them. Brass bands added to the festive air, while groups of grangers carried banners with slogans ("Labor is Honorable," "Charity and Justice to All," and "More Light to the Laboring Classes"). When the crowd assembled, Grange songs were sung, along with patriotic and religious standards. The call to order was followed by prayer (and more singing) and then by a speaker (or two). When the break for lunch was declared, families took out their picnic baskets and spent an hour or so socializing. In the afternoon, a few more speakers paid homage to agriculture and urged farmers to political action (as they saw fit); more songs might be sung; and, then, the crowds dispersed to their farms. Celebrating agriculture and instilling farmers with optimism and political energy went hand-in-hand ("Marion County," 1874; "Notes by the Way," 1874).

In what must have been a record turnout, farmers went to the polls. Despite its weak slate of candidates for state offices and lukewarm support from the Grange, the Independent Party made a strong showing. It sent 13 men to the next Indiana General Assembly (five to the Senate and eight to the House). The Democrats swept the elections as anticipated. They won all of the State offices, gained a 28-vote majority in the Indiana House, and took a one-vote margin in the State Senate (Thornbrough, 1965, p. 285). The Republican Party was spared the full brunt of popular outrage by the State Senate's staggered rotation cycle. The real victory at the polls was revealed by the fact that new men would occupy 94 of 100 seats in the Indiana House of Representatives and 24 of 25 contested seats in the State Senate (*Biographical Directory*, 1980). In the spring of 1873, Indiana's farmers had threatened to get up a movement that could "unhorse every office holder in the land" (Galbraith, 1873, p. 64). They made good on their campaign pledge.

The Aftermath: The Patrons and Politics

The 1875 session of the Indiana General Assembly became known as the Farmers' Legislature. With a Grangers' Conclave in the lower chamber setting its direction, the session differed markedly from its predecessors. The farmers' campaign showed that Hoosiers had their fill of corruption scandals, southern reconstruction, and other national issues. Action at home was wanted. In short order, the currency question (which threatened solidarity among the new office holders) was shelved, a resolution in favor of veterans' pensions adopted, and the legislature turned its attention to internal state business.

Given a sweeping mandate by the people, the new legislators embarked on an ambitious program. It included building a new State House and an Insane Asylum, revising laws related to town incorporation, the common councils, and the common school system. As promised, the temperance law was revisited. County officials' inclinations to pocket-lining were curbed with a lengthy list of fees that could be charged for services. Property taxes were slashed, assessment restored to the townships, and valuation to five-year intervals. Salaries for all state officials—including legislators—were reduced by one-third.

The Grangers' Conclave ensured that the farmer's interests were not neglected. To make capitalists aware of Indiana's natural resources, the Geological Survey received a substantial funding increase. To encourage local manufacturing, the range and privileges of corporations were expanded. A joint resolution urged Congress to impose a protective tariff on plate glass. As with the other actions, aiding Indiana's nascent glassworks industry was justifiable in the name of the farmer, since growing factories increased the non-farming population and enlarged the home market for agricultural products (*Indiana House and Senate Journals*, 1875).

The protective tariff resolution was symbolic of how the Farmers' Legislature dealt with economic issues. Aside from laws regulating grain inspection and storage (to prevent fraud and denial of access), it took no action that favored farmers' economic interests over townsmen's. It did nothing that might have interfered with the development ambitions of Indiana's towns.

Nothing revealed this better than railroad regulation. More than any other issue, the railroad companies' monopoly-power and rate-gouging had agitated farmers. Their demands for regulation were endorsed by the Indiana State Grange. At its annual meeting, held just after the 1874 elections, the State Grange advocated a moderate approach to railroad regulation, calling for proportional, not equal, rates for long- and short-hauls, as well as for permitting railroad companies a reasonable profit margin (Indiana State Grange, 1874). In its bid for farmers' votes, Indiana's Democratic Party had promised stronger action, pledging to render the railroads subservient to the people. Yet, of the handful of railroad bills introduced in the General Assembly, only one emerged from House committee, while another made it to a second reading in the Senate. Both were laid on the table.

Given farmers' demand for railroad regulation, why did granger-backed representatives fail to press the issue? A few related reasons might be suggested. Railroad representatives' argument that rate-fixing would do more harm than good was well calculated to appeal to farmers in the legislature. Defensive of personal liberty and wary of government power by nature, farmers were already inclined to oppose regulation on principle. More ammunition against regulation was provided by other Midwestern states' experience with their recently passed granger laws. Not only did the regulations prove difficult to enforce, but one of their effects was to increase the price of goods shipped long distances, particularly bulky farm implements ("Granger law-making," 1873; "The Regulation of Railroads," 1874). Railroad regulation, in short, was not only offensive to farmers' sensibilities, but bound to be counterproductive. To what extent this conviction was shared generally in 1875 is not clear. That the conviction represented the consensus among leading farmers is fairly certain; the railroad issue all but disappeared from the pages of the *Indiana Farmer*.

Dependent upon eastern city (and world) markets and manufacturers, Midwestern farmers loved railroad lines more than they hated railroad monopolies. Well aware that the railroad was "one of the great levers for opening up the agricultural and mineral resources of the West," the Indiana State Grange had urged the General Assembly to enact only laws that would "be just to the railroad interests of the country" (Indiana State Grange, 1874, p. 30) The grangers were not asking the state legislature to protect railroad corporations' property rights; that point was taken for granted. They were advocating, instead, that the legislature act to promote the people's interest in gaining access to cheap transportation. That meant building more railroad lines, creating competitive alternatives to the monopolistic (single) rail lines upon which their counties' trade with the outside world was exclusively dependent.

To advance Indiana's immediate interest in market access and its towns' long-term economic development ambitions, leading grangers wanted railroad building to receive public aid. In full agreement, the General Assembly granted township residents the authority to tax themselves to subsidize new railroad lines. At the request of 25 freeholders, a referendum on a special tax could be held. Intended to encourage the building of locally controlled short-haul routes that connected remote towns and county seats to urban centers, the approach met the approval of leading farmers and townsmen. Despite the economic depression, which lasted from 1873 to 1879, railroad building accelerated in Indiana. Nearly 1,200 miles of track were laid in the 1870s; roughly 1,600 more were added in the next decade (Barnhart & Carmony, 1954).

Building railroad lines did not solve the farmers' economic problems. Low staple crop prices, high prices on manufactured goods, a tight currency supply, and excessive freight rates continued, and agitation for railroad regulation re-gathered force. Together with the prospect of enlarging the currency supply, the demand fueled some Hoosier farmers' support of the Greenback Party. In their 1878 campaign the Greenbackers won several seats in the state legislature. Their call for reform was joined by Governor James D. Williams, Indiana's first and only Granger Governor (elected in 1876). At the next session, the General Assembly issued a joint resolution urging Congressional railroad regulation. It had little impact. Railroad regulation persisted as a force in national politics through the end of the century (Thornbrough, 1965).

When activists tried to launch the Greenback Party in conjunction with the Indiana State Grange's annual meeting they were repudiated. In their first action, the grangers issued a sharp declaration that they had "no sympathy with any past, present or future attempt that may be made by any political party or aspirant to absorb a little reflected warmth, decency or support" from association with the Patrons of Husbandry (Indiana State Grange, 1874). Subsequent political campaign seasons saw few farmer rallies hosted by local granges. Instead, farmers got reminded frequently of how the Grange had been "prostituted" by "small fry politicians" and "demagogues" in 1874 (Templin, 1875, p. 5).

After that experience, an influential element wanted a thick wall of separation between the Grange and politics. Prominent grangers who sought public office called for a respectable distance between their roles. On the Greenback ticket in 1878, Worthy Master Henley James urged subordinate grange officers "to observe the line between our duties to the Grange and our rights as citizens." If there were any doubts about propriety, they should "let the Grange have the benefit of such doubt, and travel only in the plain paths of duty" (Indiana State Grange, 1878, p. 17). Touted—or tarred—along with countless others as a farmer candidate, Henley James did not run for office as a granger. The Indiana State Grange remained steadfast in its determination to steer clear of electoral politics and campaigning.

The nonpartisan stance of the Indiana State Grange was in strict obedience with the official position of the National Order of the Patrons of Husbandry. Most grangers, however, wanted to discuss agricultural and public policy questions at their monthly neighborhood meetings. With the aid of a query box (into which members placed slips of paper to suggest topics) and committees that prepared "something in the way of lecture, essay, reading, or remark," generating discussion on the "every day employments of the farm" proved to be "no difficulty" ("Read and Discussed," 1877, p. 7). Sparking interest was even easier, surely, when the discussion turned to topics that were not strictly agricultural. When opinions were aired on national issues like the currency, railroad, and tariff—or on local issues like school laws, public salaries, dog taxes, and livestock confinement—the gavel probably did not remain silent. Nevertheless, the Indiana State Grange's traveling lecturer found that "fraternity and good fellowship reigned" inside grange meetings, despite the "noise, political turmoil, strife, and bitterness" that prevailed outside (Indiana State Grange, 1884, p. 6). As a forum for self-study, discussion, and deliberation, the Grange exercised its primary political power by educating its members about their true interests as farmers.

What Indiana's grangers failed to learn was how to convince other farmers that attending local grange meetings offered the best means for adapting to changing agricultural and political realities. The 1874 election campaign had raised farmers' expectations; after the Farmer's Legislature completed its work, the farmers' disappointment took its toll. For its disavowal of politics and its overt preference for a harmony of interests across town and country over the exclusive interests of farmers, the Indiana State Grange paid a steep price. By the early 1880s, most local granges were "reckoned as permanently gone down"; statewide membership stood at less than one-tenth of what it had been in 1875 (Indiana State Grange, 1884, p. 8). Averaging two neighborhood granges per county (some counties had none), where it existed, the Grange reached further down the social scale than the county agricultural societies of the 1850s. But, it had become remarkably similar to them in composition and outlook. The Grange was filled with only the "best material": solid, conservative, prosperous farmers who educated themselves in practical agriculture, who lived close to town, and who were fully convinced of the good that came from a flourishing home market (J. W. B., 1879, p. 7). These things represented their true interest as farmers and citizens of Indiana. Other farmers might not realize it yet, but, in time, they would.

Reflections on Civic Learning

In historical memory, the Patrons of Husbandry is remembered best as the organization farmers used in the 1870s to mobilize a popular movement for railroad regulation. The farmers' movement, however, was much more. It was a campaign waged by farmers against other economic interest groups, by common people against the elite, and by the rural West against the urban East. From a civic perspective, one that seeks to reveal forces shaping people's sense of membership in the local community, the farmers' movement was something else as well: a threat to Midwestern towns.

Farmers' outcry against railroad monopoly abuses, it is true, was not directed at nearby towns (although it was aimed at railroad agents and middlemen who lived in towns). Everyone, including farmers, knew, though, that the economic life of interior towns depended on their access to the world. Small towns, in particular, had good reason to be fearful that farmers' agitation might deter the westward flow of investment capital into their communities. The worst case scenario—strategic farmer boycotts against individual railroad lines would have been devastating to the towns along their routes. With demagogues doing their best to capitalize on farmers' discontent, who knew what farmers were capable of? Who had ever heard of farmers demanding positive action from government? The lowest taxes and the smallest reach of government into their lives was farmers' normal stance. Now, suddenly, farmers were demanding interference with the railroad, the thing towns needed most to thrive.

By itself, the political mobilization of farmers was something new and alarming. Hearing farmers grumbling about politics was typical; seeing farmers act politically was not. Farmers, as politicians reminded frequently audiences, were the steady conservative class. Had precise description been the aim, politicians would have used as adjectives the words isolated or apathetic. Occupied with their farms and families in the countryside, common farmers paid little attention to public policy. Ordinarily, political party activists were forced to take strenuous efforts (including drinks and dollars) to get out the farmers on Election Day; party principle, at best, was all farmers needed to know to vote the right way. Time and again improvement-minded farmers urged other farmers take political engagement seriously. It did little good until rate-gouging railroad companies and tax-raising legislators got farmers boiling mad.

Farmers' indifference to public policy was so profound that agricultural improvers had no qualms about invoking economic self-interest to get farmers engaged. Besides, if farmers applied their intellect, how could they help but perceive that their interest lay in encouraging economic development?

Agricultural improvers bore no special responsibility for making talk about economic interest part of the farmers' political idiom. Townsmen talked the same way. They couched everything from temperance to infrastructural programs in terms of economic self-interest and the public interest. They insisted there were no conflicts between the two kinds of interest. The civic qualifier often went unspoken: if one understood self-interest properly, long term and as a part of a broader harmony of interests with others.

Whether it entailed personal reformation or public policy, properly understood self-interest almost always pointed toward enhancing economic growth. It always included a due regard for other people in one's community. Most Midwestern farmers had never really thought of themselves as belonging to a local community of shared economic interests that incorporated their county seats. Most of their income came from supplying distant markets, and the county seat's businessmen were too inclined to take advantage of farmers when providing goods and services. For farmers to perceive themselves as living in a harmony of interests with townsmen was quite a stretch of the imagination. It was much easier to perceive what they had in common with fellow farmers as an agricultural interest. That kind of thinking pointed toward solidarity with farmers who resided in other places. From a civic perspective, first and foremost, farmers' loyalties belonged to their place-based communities, to their immediate township and county.

The antimonopoly and anti-town sentiment of the farmers' movement obscures the fact, but the Grange intended to teach farmers to be more civic-minded. Deliberately, the organization's membership was not restricted to farmers only; non-farmers who were interested in agricultural pursuits were allowed to join. Fostering a harmony of interests through agricultural improvement and economic development was a core tenet of Grange teaching. Perhaps leading grangers were naïve to think they could organize farmers without stirring-up a farmer's campaign against all other economic interests (although they could not have predicted the financial panic of 1873). Yet, moderate reform that benefitted town and country was what they wanted. Ultimately, that was what Indiana's farmers got from the Farmers' Legislature of 1875.

Some of the credit for this was due to prominent grangers' determination to prevent a farmer's party from forming. Some credit was due to the partisan spirit prevailing among farmers that was deplored constantly by agricultural improvers. Amid the calls for purifying and restoring politics to true principles, farmers returned to their traditional party, the Democratic Party. Inside the Democratic Party, since leading townsmen and farmers controlled affairs, the nominating process yielded the kind of candidates respectable grangers wanted in public office. Well versed in the proper civic perspective, they made sure that political power was not used to enact policies that conflicted with the towns' economic development ambitions. As such, involvement in the farmers' movement and Grange of the early 1870s gave Indiana's farmers an important and long-lasting civics lesson: if farmers' demands interfered with the prospects of Indiana's towns, the political process offered little hope. It was a powerful lesson from political experience, one that could not be taught in schools.

References

A. C. (1873). [Untitled letter]. *North Western Farmer, 8*(3), 66.

Barnhart, J. D., & Carmony, D. F. (1954). *Indiana: From frontier to industrial commonwealth* (Vol. 2). New York: Lewis Historical Publishing Company.

Barnes, W. D. (1967). Oliver Hudson Kelly and the genesis of the Grange: A reappraisal. *Agricultural History, 41*(3), 229–242.

Billingsley, J. J. W. (1873). Grange mass meeting in Tipton County. *Indiana Farmer, 8*(2), 6–7.

Biographical directory of the Indiana General Assembly (Vol. 1). (1980). Indianapolis, IN: Indiana Historical Bureau.

Buck, S. J. (1921). *The agrarian crusade: A chronicle of the farmer in politics.* New Haven, CT: Yale University Press.

Capital and Labor. (1874). *Indiana Farmer, 9*(8), 4.

Chase, B. S. (1873). Monroe County farmers' club. *North Western Farmer, 8*(7), 159.

Collett, W. (1874). Address before Lagrange County council. *Indiana Farmer, 9*(15), 1.

Congress. (1874). *Indiana Farmer, 9*(24), 4.

Conner, J. B. (1873). Granges in Newton County. *North Western Farmer, 8*(4), 87.

Deal, S. (1873). Co-operation. *North Western Farmer, 8*(2), 42.

Demaree, A. L. (1941/1974). *The American agricultural press, 1819–1860.* Philadelphia, PA: Porcupine Press.

Doty, D. (1873). Club in Pike County. *North Western Farmer, 8*(5), 109.

Farmers should exert their influence. (1873). *North Western Farmer, 8*(5), 113.

Farmers should organize clubs and societies. (1873). *North Western Farmer, 8*(1), 19–20.

From an objector to the order. (1873). *North Western Farmer, 8*(3), 65.

Galbraith, M. V. (1873). Objects of the order. *North Western Farmer 8*(3), 64.

Grange pic-nic—Cartersburg, Hendricks County. (1874). *Indiana Farmer, 9*(30), 4.

Granger, A. (1874). Patrons and politics. *Indiana Farmer, 9*(20), 5.

Granger law-making against railways. (1873, August 14). *Dearborn Independent.*

The Grangers in motion. (1874, August 13). *Aurora Farmer and Mechanic.*

Greencastle Banner. (1873). [Untitled reprint]. *Indiana Farmer, 8*(4), 5.

Harvey, A. C. (1873). Grange pic-nic. *North Western Farmer, 8*(8), 181.

Henry, W. E. (1902). *State platforms of the two dominant political parties in Indiana, 1850–1900.* Indianapolis, IN: Author.

Higher ground. (1873). *Indiana Farmer, 8*(14), 8.

The independent convention. (1874). *Indiana Farmer, 9*(24), 4.

The independent ticket of Marion County. (1874). *Indiana Farmer, 9*(25), 5.

Indiana House and Senate Journals. (1875). 48th Regular and Special Session.

Indiana State Grange. (1873). *Third Annual Proceedings.* Indianapolis, IN: Author.

Indiana State Grange. (1874). *Fourth Annual Proceedings.* Indianapolis, IN: Author.

Indiana State Grange. (1875–1876). *Fifth Annual Proceedings.* Indianapolis, IN: Author.

Indiana State Grange. (1878). *Eighth Annual Proceedings.* Indianapolis, IN: Author.

Indiana State Grange. (1884). *Fourteenth Annual Proceedings.* Indianapolis, IN: Author.

Indiana State Grange. (1889). *Nineteenth Annual Proceedings.* Indianapolis, IN: Author.

J. W. B. (1879). [Untitled letter]. *Indiana Farmer, 14*(6), 7.

Launching the craft. (1874, July 16). *Indianapolis Sentinel.*

Mace. (1874). Grange and politics. *Indiana Farmer, 9*(20), 1.

Marion County patrons picnic. (1874). *Indiana Farmer, 9*(30), 4.

Mass meeting of the grangers at Remington, Jasper County. (1873). *Indiana Farmer, 8*(1), 5.

McCoy, A. B. (1874). Politics and patrons. *Indiana Farmer, 9*(22), 5.

Mills, S. (1871). Decatur Township agricultural society. *North Western Farmer, 6*(12), 60.

Nordin, D. S. (1974). *Rich harvest: A history of the Grange, 1867–1900.* Jackson: University Press of Mississippi.

Notes by the way. (1874). *Indiana Farmer, 9*(30), 4.

Parties. (1874). *Indiana Farmer, 9*(5), 4.

The patrons and politics. (1873a). *Indiana Farmer, 8*(5), 5–6.

The patrons and politics. (1873b). *Indiana Farmer, 8*(7), 159.

Progress of the order. (1873). *Indiana Farmer, 8*(12), 10.

Ratliff, J. C. (1873). Wayne County. *North Western Farmer, 8*(6), 128.

Read and discussed. (1877). *Indiana Farmer, 12*(19), 7.

Reform: The independent convention. (1874). *Indiana Farmer, 9*(23), 1.

The regulation of railroads. (1874). *Indiana Farmer, 9*(19), 4.

Stevenson, A. C. (1873). Remarks before the Clay Township farmers' club. *Indiana Farmer, 8*(2), 2.

Templin, L. J. (1873). Patrons of Husbandry. *Indiana Farmer, 8*(7), 3.

Templin, L. J. (1874). Pickings for the patrons. *Indiana Farmer, 9*(1), 2.

Templin, L. J. (1875). The grange prostituted. *Indiana Farmer, 10*(29), 5.

The tenth of June convention. (1874). *Indiana Farmer, 9*(20), 4.

Thornbrough, E. L. (1965). *Indiana in the Civil War era.* Indianapolis: Indiana Historical Bureau and Indiana History Society.

To the county and district voters. (1874, May 9). *Indiana Farmer, 9*(18), 4.

Wennick, J. H. (1873). Shelby County organizing. *North Western Farmer, 8*(7), 159.

What does this movement of farmers mean. (1873). *Indiana Farmer, 8*(11), 8.

Woods, T. A. (1991). *Knights of the plow: Oliver H. Kelley and the origins of the Grange in republican ideology.* Ames: Iowa State University Press.

The word on the approaching convention. (1874, July 13). *Indianapolis Sentinel.*

6

Between Town and Country

The Grange and Economic Cooperation

"A Granger's Dream"

A Granger dreamed that he died and went straight to the spirit world. The books were opened, and he was asked: "did you ever belong to any secret societies?" To this he replied, "I did, to the Grange." "Then, sir, you cannot be admitted." The granger then went to the bottomless pit, where the same question was asked. Again he was told to depart. After he had gone a little way he was accosted by the homely ruler of the pit, who made him a proposition. "Stranger," said Nick, "they do not want you in Heaven and I will not admit you here. But I will do this for you. I will sell you two hundred barrels of brimstone for cash, ten per cent off, and you can start a little hell of your own, with no agents or middlemen." ("A Granger's Dream," 1874)

On the necessity of preserving their economic independence, Midwestern farmers were in solid agreement. Being an independent farmer ultimately meant one thing: free and clear ownership of the farm. In the transitional era when farmers engaged both in subsistence agriculture and commercial production, the route to independence came in two broadly defined forms. Farmers in the countryside relied heavily upon the farm for survival

Civic Learning through Agricultural Improvement, pages 123–150
Copyright © 2011 by Information Age Publishing

necessities and earned cash income from shipping grains and meat to distant urban markets. A much smaller group of farmers who lived closer to town sold staple crops to distant urban markets and supplied lesser produce (dairy and poultry products, vegetables, and fruits) to nearby villages and cities. The remaining farmers tried to make the farm pay by whatever means seemed feasible given their location and circumstances.

In the 1870s, Midwestern farmers needed cash income desperately to pay off debts from land and machinery purchases made during the boom years of the Civil War. Agricultural improvement's main prescription was to diversify and expand subsistence practices so as to supply town-dwellers' tables. Few farmers could do that, realistically. With a small town-dwelling population, the home market demand for foodstuffs was sharply limited. Town-dwellers' kept their own gardens, fruit trees, and small livestock; when economic depression hit, they tightened their belts by growing more of what they needed. Farmers were left with only one option: trade more efficiently with distant markets.

The typical farmer was no economist, but he knew what had to be done. As a producer, he needed to reduce costs; as a consumer he needed to obtain savings. Foremost among expenses were the extortionate fees, markups, and interest rates exacted by local town-dwelling middlemen—warehouse operators, commissioned agents, merchants, and sales agents for farm machinery companies—who handled his business with the outside world. Down with the middleman! Direct purchasing and selling became watchwords. Clubbing together, farmers bypassed local middlemen and convinced distant suppliers to strike better deals using economies of scale and cash-on-delivery, rather than credit. Where the use of middlemen could not be avoided, such as selling farm produce to distant markets, grangers appointed their own. By combining their efforts, farmers hoped to eliminate the need for some middlemen and to teach the remaining middlemen to do business on reasonable terms. Securing their independence required farmers to assert their economic independence of the town.

Farmer-dollars carried clout. Confronted with boycotts and outside competitive pressure, local merchants and middlemen were compelled to lower their rates. Yet, at the peak of their strength, the Indiana State Grange abolished farmers' most formidable weapon, its State Purchasing Agency. With it, grangers had an institution that pooled their buying power across the state, into an economy of scale against which county merchants and middlemen could not compete. Deprived of the State Purchasing Agency, grangers had limited bargaining leverage, at best, in their local communities. Merchants and middlemen went back into their usurious ways. Farmers' demands to restore the State Purchasing Agency went unheeded by the State Grange. Economic cooperation among the farmers disintegrated.

Why did the State Grange sound the retreat? The case of Indiana suggests that the granger's dream began to look like a citizen's nightmare. Compelled to strike a deal with the devil (distant big city suppliers) against local merchants and middlemen, when leading grangers witnessed the effects of farmers' combined economic mite on their towns' economic well-being they found that they no longer wanted a little hell of their own. Instead, they wanted their farms annexed into a town-based Heaven. Changing course, their preferred version of economic cooperation among farmers was local in scale and limited in scope, with the thrust of its efforts devoted to educating farmers in practical economics rather than to reforming business practices of townsmen. Through the abolition of the State Purchasing Agency, Indiana's most prominent farmers revealed their civic-mindedness. They preferred striving for a harmony of interests between townsmen and farmers over emphasizing their conflicting interests. They preferred, too, the mutual dependence of town and country over the independence of all farmers.

Combining Farmers into an Economic Power

Midwestern farmers had never needed mutual protection so badly. The plunge in prices for wheat, corn, and hogs to about one-half of what their value had been at the close of the Civil War was a bitter pill (Thornbrough, 1965). What made it worse was that non-farmers were living large. How their profits were obtained was easy to discern. Pork packers and grain buyers met in convention to fix prices. Implement manufacturers sold their machines at triple the actual cost of production; by fixed agreement, their local sales agents got a commission of 25 percent or more. Local storekeepers were even worse for tempting farmers with easy credit, then tacking-on high interest rates. Farmers had always complained about town-dwellers' corruption—their inclinations to usury, dishonest dealing, and luxurious living—but this degree of corruption was unprecedented.

Everywhere the story was the same: whoever the farmer had to deal with was part of a larger combination, a network of creditors, suppliers, and purchasers that worked behind the scene to fix prices at artificially high levels. At a bargaining disadvantage, the isolated farmer had no choice but "to submit to their exorbitant rates, and take for their products just such prices as they [purchasers of farm produce] see fit to give" (Doty, 1873, p. 109). If men in other trades could combine to advance their interests, then the "spirit of organization" was the "first thing for farmers to learn" (Galbraith,

1873, p. 64). Hard experience proved how much they needed mutual protection; farmers were "ready and willing to put [their] shoulders to the wheel and help push along this great work" (Thorntown, 1873, p. 5).

Midwestern grangers applied economic cooperation to virtually every point of encounter with the cash economy. Granger-owned grain elevators, flouring mills, slaughterhouses, and machinery factories received high profile coverage in the nation's newspapers, but obtaining cheaper clothing, goods, and groceries commanded the bulk of farmers' attention. Regardless of the goods, the cooperative principles were the same: pool orders to enhance purchasing power; deal on a cash-only basis; keep terms confidential; eliminate middlemen where possible; and if middlemen were indispensable, contract with friends of the farmer (Cerny, 1963). In principle, the business feature, as economic cooperation was termed, of the Grange was simple. As long as farmers stood united they had a powerful tool for bargaining leverage in the commercial economy.

Traditionally, farmers had done their own buying and selling. As often as not, the two sides of commerce went hand-in-hand on farmers' occasional trips into town. Granger cooperation built upon this practice. At the county council, orders for goods were compiled and a member was appointed to place the order. Another was appointed to travel from grange to grange loading farm products into his wagon. At the county seat he left the produce and picked up the goods. He dealt only with pre-approved places of trade, for at their meetings, grangers compared price lists, from local and distant points, to determine what they thought were reasonable prices (Thomas, 1873). Deliberating on terms with purchasers of farm produce, bargaining with merchants, and divvying-up the shopping was cumbersome, but it worked. Purchasing from suppliers who "made reasonable concessions" and sparing "the middleman the trouble" of handling orders, farmers saved 20 to 50 percent (and more on some items) on their purchases (Shiloh, 1875, p. 5).

The work of the county councils convinced farmers that economic cooperation was effective. They pushed it to the next level. By the close of 1874, in more than two dozen states, State Grange Purchasing Agents were appointed (Cerny, 1963). Local experimentation and the diffusion of ideas gave grangers' economic cooperation its real potential. The concern that farmers might have discovered the right formula, set off alarms from St. Louis to Boston. Heavy-hitters in public opinion-making—notably, E. L. Godkin of *The Nation* and Charles Francis Adams for *The North American Review*—went to work (Nordin, 1974). Funds flowed to finance new papers in the West to denounce "the *secret* intermeddling with established commercial customs" carried out by grangers ("The Grange," 1874, p. 4). Sensibly, Midwestern publishers reserved the "misrepresentations" of granger activities for the daily edition, while pointing "with *great pride* to their column of 'choice agricultural reading matter'" in the weekly edition (A Patron, 1875, p. 5). Publishers knew who read which papers: for the town, an anti-granger campaign; for the country, agricultural improvement.

Financial support came from the East; the press and pulpit lent their services. The encouragement was unnecessary. If Midwestern farmers could band together "to throw off all yokes and look after our own interests," townsmen could organize to keep on the yokes and protect their interests, too (Harvey, 1873, p. 160). Merchants and middlemen in towns closed ranks and refused to deal with the local grange or its members. Bankers threatened to foreclose farm mortgages and businessmen to bring debt-collection lawsuits against farmers who showed an interest in the Grange (Clay, 1876; "Futile Opposition," 1876). Persecuting potential grange-joiners was "bad policy." Even the dullest wit could perceive that the intimidation tactics would serve only to arouse farmers' indignation and "hurry them in in greater numbers" (R. H. K., 1873, p. 160).

The gap between town and country was widening and resolve hardening. Yet, all the while perfecting their cooperative techniques, grangers insisted that farmers were not waging war against merchants, middlemen, or nearby towns. Rather, they were taking action as a means of self-defense, as a form of mutual protection against extortionate and abusive business practices. Farmers opposed only the needless hand-changing and excessive profit-taking that occurred between farm and market (and vice versa). What they wanted was fair treatment on legitimate business terms. Grangers were "willing to travel with the manufacturers and dealers of the country," if they were willing to go the farmers' way "half the time" (Miller, 1875, p. 5). Did not the Grange offer a good deal? The agent who handled grangers' outgoing grain and livestock got guaranteed trade and a reasonable commission. On the incoming side, the merchant who declared himself a friend of the farmer was assured steady sales and cash payment. Fair dealing and legitimate profits only, those were the grangers' conditions.

The town elite might snub their lesser counterparts for letting farmers dictate the terms of trade, but hard times were hard times. Merchants and middlemen began joining the rush of farmers into the granges. A few may have planted pumpkins to establish their agricultural bona fides, but they need not have bothered. Were not they all interested in agricultural pursuits? That was what the Grange constitution required. Never keen on the business side of farming, grangers appointed their new friends to handle

it. From farm to market, and factory to farm, a Grange-affiliated infrastructure of suppliers, handlers, and purchasers took shape. Non-farmers of all
stripes began "asking for terms . . . and admission to our gates to the field of
reform; for all see and acknowledge the good effects of our order" (Ross,
1876, p. 5).

Giving the leg up to competitors produced changes (if temporary) in
the economic behavior of other merchants and middlemen. Big city retailers on all sides were offering "special bargains" to farmer clubs and grange
associations (Ayers, 1874, p. 4). By demanding direct trade, farmers sent a
goodly portion of the "army of drones and go-betweens" back to the plow.
Made all the wiser from witnessing their experience, remaining middlemen
began dealing "directly with the membership on honorable and generous
terms" ("The Grange Organization," 1874, p. 4). By the mid-1870s, farmers could obtain most goods for about one-half of the 1870 price ("The
Grange," 1877).

The most vivid testimony of granger strength came after a "hair scrap-
ing conclave" of Midwestern pork-packers fixed the hog price in the three-
dollar range ("Why Don't Somebody Howl?" 1874). The battle cry for an
organized response went out the next season. Urged to hold their hogs off
the market and to pack their own pork if necessary, the farmers did, until
the pork packers agreed to pay the price grangers insisted "their pork was
honestly worth." Even the *New York Times* conceded the point: "farmers can
combine on a larger and grander scale than any other class, when it be-
comes necessary" ("What Has Been Done," 1876). The campaign against
the pork packers seems to be the only instance in which farmers managed
to name their price on a regional scale. Still, at the time it was a striking
object lesson, powerful proof that farmers could organize for mutual protection of their interest.

Concentration of Cooperation: Competing Priorities
and Complications

Bringing the pork packers to heel was heady stuff. So, for that matter, was
seeing a well-dressed merchant come to the grange council hat-in-hand
seeking terms. The year 1875 was the pivotal moment. The political excitement of the election campaign was over; grangers could redouble their
efforts in economic cooperation. The Indiana State Grange had a war chest
of $25,000 from the dues paid by its 65,000 members. Rightly, with appre-
hension and anticipation, townsmen and farmers wondered: what would
the grangers do with their money and numbers?

What remained to be done? Prices had fallen. Opposition from local merchants and middlemen was on the wane. Had the grangers achieved their aims? A few thought so, but, in general, the prevailing opinion favored continued cooperative action. Some prominent farm implement manufacturers still refused to abandon the local agent system in favor of direct trade. On principle—and to prevent their example from encouraging backsliding among others—they would have to be brought to terms. Rural counties were still plagued by a surplus of merchants. Their numbers would have to be reduced and their credit system extinguished. Why not bring the power of the Grange to bear upon all manufacturers, middlemen, and merchants who continued to take advantage of the farmer? That was the path to lasting reform. Had the considerations involved been strictly economic, the farmers might have succeeded. How best to save money furnished trouble enough. That issue's entanglement with civic considerations about how grangers' cooperation affected nearby towns ensured failure.

By itself, the middleman problem pulled grangers in different directions. Middlemen operated on three distinct tracks: sending farm produce (especially grain) to distant urban markets; getting the means of production from suppliers, particularly farm implements; and, purchasing consumer goods and groceries. All farmers shipped grain and meat to distant urban markets. But coordinated statewide export efforts were not practicable in Indiana, since farmers in different parts of the state sent produce to different markets, such as Chicago, New Orleans, and Cincinnati. Sensibly, the State Grange left "bulking up the product of the soil and selling wholesale to the parties who pay the highest price" to local determination (Indiana State Grange, 1874, p. 31). Coordinated statewide action could be taken against the two other sets of middlemen. Implement manufacturers' local agents were given territorial monopoly privileges and fixed commissions that raised farmers' operating costs. Practically speaking, local merchants also possessed a near-monopoly privilege, since they controlled farmers' access to consumer goods and groceries. Farmers wanted to save money on both sides, as producers and consumers.

When Hoosier Patrons emerged from the State Grange's annual convention at the end of 1874, they set out to work both fronts. Grangers had proven that they could dispense "with the old system of purchasing through middlemen." Their county councils no longer had any difficulty "establishing direct trade with manufacturers and wholesale men" on favorable terms (Indiana State Grange, 1874, p. 12). The grangers' efforts, however, were too independent and too local. Organized by separate county councils, grangers' combined force (in each county) exerted only a limited pressure on distant suppliers. The county councils' actions, moreover, undercut each other, since, working independently, they reached agreements with different wholesalers at different rates. The work of the county councils needed to be consolidated on a statewide basis.

A State Purchasing Agency had been operating in Indiana since the autumn of 1873. In *Indiana Farmer* editor J. G. Kingsbury, the State Grange found the right man to put it into operation. Corresponding with county councils and negotiating with Indianapolis-based wholesalers, Kingsbury got the orders rolling into Indianapolis and the farm implements, sewing machines, and dry goods rolling out. The volume of correspondence compelled Kingsbury to hire a clerk and an assistant editor; the commerce required additional rooms and a man to tend to the displays and orders. After several months, with one of the fastest-growing farm papers in the Midwest and a farm to supervise, Kingsbury turned the State Purchasing Agency over to Alpheus Tyner ("Meeting of the Executive Committee," 1873; "State Purchasing Agency," 1874).

A proven success at handling the business affairs of the Shelby County Council, Alpheus Tyner committed himself on a full-time basis to the State Purchasing Agency in the summer of 1873. He distributed more than 100,000 circulars, price lists, and pamphlets to grangers, and by year's end the "good results" of his broadcasting were apparent. More than $310,000 in commerce passed through his office. The State Grange Agent was efficient. His average markup was less than 2 percent per transaction and expenses totaled less than $6,000. Sustaining the enterprise cost each Hoosier granger less than nine cents (Indiana State Grange, 1875–1876, pp. 19–20). Impressed by Tyler's efforts, the State Grange upheld the State Purchasing Agency as the "proper channel" to furnish grangers "all articles needed for the cultivation of the farm, and for the household" (Indiana State Grange, 1874, pp. 35–36).

As is typical of large delegate bodies, the State Grange passed resolutions and left the operating details to its Executive Committee. Their charge was straight forward: to devise a system to pool orders among the granges of each township, then again at the county-level, for forwarding to the State Agent. For farm implements, that is exactly what the Executive Committee stipulated. For goods and groceries, it insisted that farmers' cooperation should be restricted to the smallest, most local, unit. In each rural neighborhood, grangers were to compile orders, and, with cash in hand, "try the nearest merchant." The State Purchasing Agency was to be used as a last resort, only for goods and groceries unavailable locally. To the fullest extent possible, Alpheus Tyner was to "give his attention to the State agency as it was intended," as a medium for making terms with agricultural implement manufacturers (Newsom & Mitchell, 1874).

The Executive Committee's plan of operations was a deliberate attempt to purge the State Purchasing Agency of half its usefulness. Robert Mitchell, Secretary of the Executive Committee, laid out their reasoning. It was neither feasible nor desirable for Tyner's Agency to handle grangers' commerce in goods and groceries. The volume of orders from all the counties of the State would overburden the Agency, and using Indianapolis as a way station (between distant points and the final destination within Indiana) would increase freight charges. That part of the argument might have seemed reasonable to some farmers; the rest of it would not. Mitchell insisted that farmers could get reasonable prices from local merchants, if they were approached "in a business spirit." If farmers failed to do this and transferred their commerce to Indianapolis via the State Purchasing Agency, their action would "blot out of existence all the thriving towns, reduce the value of [farmers'] lands, and local markets." Grangers, therefore, should "keep the business [at home] where it properly belongs." Coming from a man who got his start in life "following the plow at 50 cents a day," it was an oddly civic line of reasoning (Mitchell, 1875, p. 5).

Most farmers did not live near thriving market towns and their chances of persuading local merchants to meet them in a business spirit were slim: the more remote and smaller the town, the fewer the merchants and the tighter their grip on local commerce. The Executive Committee knew this. They also knew that if a town's location was unfavorable its livelihood was precarious. Pooling purchases at the county- and state-level gave farmers direct access to outside suppliers against whom local merchants could not compete; restricting pooling to the rural neighborhood preserved a place for local merchants in the circular flow of commerce. One granger wanted to know if the Executive Committee was deliberately "throwing everything back where it started, in the hands of middlemen" (Jackson, 1875, p. 5). Another wondered if the "Worthy Committee" was incapable of comprehending that on every transaction that local merchants conducted farmers were "paying a double set of clerks" (Granger, 1875, p. 5).

Hoosier farmers were not willing to give up their savings for the sake of supporting the small town economy. They had tried "purchasing at home, from local dealers...long enough to know" what that had to offer and did not want "any advancing backwards" (J. K., 1875, p. 5). The State Grange voted Alpheus Tyner a salary and a capital fund on "the understanding that the Order might purchase *anything* through the agency that suited them" (Grant, 1875, p. 5). Alongside the hay rakes and cultivators, barrels of sugar, kegs of nails, and a sundry of goods continued to pass through Tyner's office. Under pressure, the Executive Committee put forward a "new method" for ordering goods. It was identical to the method for ordering farm implements. Orders from subordinate granges were to be pooled at the county-level "promptly on the first of each month" for forwarding to the State Agent. A new State Council of County Purchasing Agents was created to work out arrangements with Alpheus Tyner and suppliers ("Proceedings of the Executive Committee," 1875, p. 4).

The *Indiana Farmer* thought the grangers' business feature stood on "the threshold of the most complete success" ("The New Method for Business," 1875, p. 5). Several months of close coordination between the county councils and the State Agency seemed to bear out the prediction. An August report claimed statewide purchasing of approximately one million dollars per month through the cooperative network created by the State Agency and county councils ("State Agents," 1875, p. 5). The savings notwithstanding, farmers' discontent grew. Grange ordering and purchasing policy were conducted on a cash only basis. Bulking orders once per month at the county level was difficult for poorer farmers who lacked the cash to order a supply of goods that would carry a family for several weeks. Other grangers raised principled objections to the new method. The State Purchasing Agency used a system of county agents. Was not the Grange creating its own tier of middlemen? Why bother with the State and County Agency purchasing circuit at all, if orders could be sent direct to suppliers and deliveries could be made locally (Patron, 1875, p. 5)?

By August, farmers' efforts to grow their own locally based cooperative projects were well underway. Fearful of losing access to cheap goods and groceries when the Executive Committee tried to reduce the State Purchasing Agency's role to implement acquisition, farmers had redoubled local efforts. With willing townsmen nearby, it was not difficult for grangers to create cooperative stores. In Howard County, for example, $1000 worth of subscriptions from subordinate granges launched a supply house that handled $30,000 in sales in its first year and kept in stock "anything needed . . . at prices so low as to astonish the natives" ("Howard County Grange," 1876, p. 5). Grangers also had the physical facilities. Two-story halls were replacing the schoolhouse as the grangers' meeting place; often, they doubled as cooperative stores or warehouses. The Executive Committee accelerated the trend toward local dispersion when it reluctantly endorsed cooperation at the county level. Notably, the county-level cooperation that received official sanction was conducted on the joint stock principle; this was cooperation among individuals, not cooperation among organized groups of farmers (Newsom & Mitchell, 1874). With three types of cooperative activity operating in local communities—neighborhood grange, county council, and joint stock—a broadened pool of grangers could perceive the merits of spreading the farmers' purchasing power all over the State. Controversy over the question of abolishing the grocery and supply function of the State Purchasing Agency would occupy the State Grange's annual meeting in December (Watchman, 1875).

Using the State Purchasing Agency to obtain small goods and groceries was not the same thing as using it for farm machinery. There was no legitimate civic reason why (most) Hoosier Patrons should oppose consolidated cooperation against nationally recognized implement manufacturers who refused to abandon their local agents. With two prominent exceptions (with one of which, the Studebaker Wagon Works, grangers seemed to have no difficulty), none of them were located in Indiana. Many of the implement makers who were willing to come to terms with Alpheus Tyner were from Indiana or adjacent states. With an eye to encouraging the growth of local industry, the State Grange convention instructed members to give these manufacturers "hearty support" via Tyner's office (Indiana State Grange, 1874, p. 26). The Executive Committee was equally enthusiastic. Indeed, why could not grangers establish a showroom in each county to bring the implements on Tyner's list to the attention of local farmers? Eliminating the middleman and encouraging home manufacturing: granger principles and the civic imperative fit together perfectly (Newsom & Mitchell, 1874).

Convincing farmers to see matters that way was not readily achieved. To the typical farmer, boots were boots and sugar was sugar. The same could not be said of farm implements. From attending competitive demonstration trials at county fairs and grange events, farmers knew different brands by reputation and performance. In an age of rapid technological innovation, different brands displayed distinctive design features. These features determined not simply which brand of, say, seed drill, a farmer would purchase, but whether he would purchase a seed drill at all. Critics of the Grange often asserted that the organization's purchasing agencies offered only inferior machinery. The claim was not quite true, for Alpheus Tyner's line of implements won 27 first-place and 11 second-place premiums at the 1875 State Fair ("The Grange at the State Fair," 1875, p. 5). Nevertheless, even the State Agent had to admit that many farmers showed a "lack of interest" in his implements (Indiana State Grange, 1875–1876, p. 22). The gap in quality, as well as distinctive features, between good locally made implements and the best nationally distributed implements was too large to change farmers' preferences.

No farmer could do without a plow. The plow farmers across the nation wanted most in 1875 was the Oliver Chilled Plow. Made in South Bend, Indiana, it was the hands-down winner at several states' State Fair competitions in 1874 ("The Oliver Chilled Plow," 1874, p. 4). Its chilling process for making steel plowshare points, adjustable dimensions, and interchangeable system of parts gave the Oliver Chilled Plow its status as the lightest-pulling, most-durable, and best all-purpose plow. Priced at less than fifteen dollars, it was the cheapest plow on the market, too (South Bend Iron Works, 1874, p. 2). Only one objectionable feature marred the Chilled Plow's perfection. James Oliver refused to disband his network of local sales agents.

Early in 1875 Alpheus Tyner placed the matter before the Executive Committee. The call went out for a boycott ("Proceedings of the State Council," 1875). If the flurry of letters submitted to the *Indiana Farmer* is used as an index, Hoosiers' outrage at James Oliver surpassed their anger at the tax-raising, salary-grabbing General Assembly two years earlier. Best and cheapest notwithstanding, the true granger would not "sigh after an Oliver Chilled Plow" until it could be purchased without the "need of a great class of middle-men to stand between us and the manufacturers" (S. B. B., 1875, p. 4). Inundated with resolutions, the *Indiana Farmer* editor declared the sentiment to be "almost or quite universal" and resorted to listing only the names of the sponsoring granges. Correspondence usually tailed-off during farmers' busiest months and letters on the boycott continued into the summer ("Patrons of Husbandry," 1875, p. 5).

Anti-Oliver sentiment among grangers—much less among all farmers—was not as universal as the dozens of resolutions posted in the *Indiana Farmer* indicated. Who, some grangers wanted to know, did the Executive Committee think they were to "assume guardianship over the members of the Granges, about their business, and what plows they shall and shall not purchase?" What permitted them to exercise such "dictation and tyranny" over how farmers spent their own hard-earned dollars (R. S., 1875, p. 5)? "Hundreds of Patrons [were] using the Chilled Plow, because they like it the best" (A. L. D., 1875). Leaders of local grange clubs bought Oliver Chilled Plows and county councils included them in demonstration trials (Comstock, 1875). The *Indiana Farmer* carried advertisements extolling the Oliver Plow's virtues. The grange-sponsored boycott itself, though, was the best possible endorsement. By debating the Oliver Plow endlessly in the agricultural papers, grangers provided "first-class advertisements" for the South Bend Iron Works, since farm papers around the nation reprinted each other's articles (Fisher, 1875, p. 5). The boycott was an utter failure. Despite the economic depression, James Oliver could not keep pace with the demand for his plows.

The Oliver Plow boycott carried a steep price for the Indiana State Grange. It made common farmers more aware of how different their interests were from prominent farmers. Perhaps it was the Grange leadership (not common farmers) that was getting carried away? James Oliver capped his agents' commissions at 15 percent and offered the best and cheapest plow on the market (H., 1875, p. 5). Why should farmers boycott him? Had not the Grange leadership declared—on principle and to protect townsmen's commercial interests—that grangers should oppose only abusive and needless middlemen? With regard to goods and groceries, the Executive Committee told farmers to be reasonable, to purchase from "home merchants" for the sake of "building up" the home economy? Why, then, should farmers refuse to buy plows from "home agents" who sold the Oliver Plow (Forward, 1875, p. 5)? The South Bend Iron Works was located in Indiana; buying its plows, therefore, was one way for farmers to support home manufacturing. Consistency thou art a jewel.

The logical contortions and divided opinions only got worse. In their arguments against cooperative stores, leading grangers had insisted that farmers had no business engaging in any work other than farming. Now some grangers were entertaining the idea of branching-out into the manufacturing pursuits. The Indiana State Grange had received a "magnificent offer" of land and buildings (estimated at a $50,000 value) to establish an implement factory at Indianapolis. Some "experienced manufacturers" from Pennsylvania were willing to staff it. Would Hoosier Patrons raise another $50,000 in stock subscriptions? If they did, as majority shareholders, they could oversee operations by electing the board of directors of the proposed "Patrons Manufacturing Association." Made, shrewdly enough, between the fall harvest and the annual State Grange meeting in December, the offer was good for 30 days. A "number of Patrons from different parts of the State" met in Indianapolis. A circular, complete with articles of association and subscription-blanks went out to the subordinate granges ("The Opportunity Offered," 1875, p. 4).

From their experience with the failed Oliver Plow boycott, apparently, some grangers concluded that they had been misdirecting their energies. The "fault," it seemed, was "not *all* with middle men." Some manufacturers simply could not tolerate the thought of being "dictated to by a set of blockhead farmers" in their business affairs (J. S. D., 1875, p. 5). So be it: there were plenty of farmers with money hoarded away to purchase $25 shares to support a farmer-owned implement factory. Why should not farmers put their "idle dollars, here and there, together" to work for their combined interest? If farmers did, the proposed implement manufactory would serve as "a solid token of our strength and ability to do as well as talk." It would not be long before the recalcitrant implement manufacturers realized that they could not simply ignore farmers' demands for reform (S. G., 1875, p. 5).

The call to support the Patrons Manufacturing Association echoed the farmers' movement. It was not being made on behalf of the farmer's independence, however. The campaign was a call for Indiana's leading farmers and non-farmers to join hands to promote economic development. Building a large scale implement factory in Indiana would halt the "suicidal" policy of sending "into the laps of our sister States" the two million dollars spent annually on farm implements by Hoosier farmers (Billingsley, 1875, p. 5). Directly, it would employ dozens of workmen, but indirectly, by using Indiana hard wood, Indiana coal, and Indiana iron, it would employ hundreds more. In this way, the Patrons Manufacturing Association would foster a sizable and steady home market for farmers' produce. If little else, its Indianapolis location would save farmers the shipping costs on their machinery purchases (Indiana State Grange, 1875–1876).

Despite these benefits, the Patrons Manufacturing Association was unlikely to gain much support among farmers. Most farmers had no use for large implements (other than plows) and no idle dollars to invest in manufacturing, only the wealthiest farmers did. The prospect of securing widespread support, even among this small group, was dim. Location mattered, for Indiana's elite farmers were also the state's most civic-minded farmers. Wealthy farmers living near Indianapolis could give the listener an earful of why it was a good idea. Men from other regions, in turn, could put to good use Robert Mitchell's argument against building up the commerce of Indianapolis.

Regardless of location, a number of farmers were bound to oppose the Patrons Manufacturing Association on principle. Concerned with preserving the "unity of interests" among farmers, they wanted the Grange to be an organization only for farmers. This meant keeping all joint stock ventures at bay, including cooperative stores. Involving their granges in non-farming enterprises would undermine farmers' solidarity against town-based merchants and middlemen. Farmers' interests and the interests of those with whom they did business were irreconcilable: in buying and selling, one party always stood to gain at the other's expense. Those "conflicting interests" could not "be harmonized by originating them in the Grange." Men who imagined that "lambs of their breeding, will lie meekly" would do well to recall that lambs grow "into old bucks with heads as hard and hearts as unrelenting, as those who have gone before." As the joint stock enterprises got established, their investors (wealthy farmers) would protect their investment and profits, if need be, against the interests of other (common) farmers. An institution for mutual protection could not be sustained unless its members "were held together by a unity of interests" (J. C. A., 1876).

A growing number of grangers opposed the Patrons Manufacturing Association for a different reason. They were "out of patience" with co-operative schemes of any kind. Already, the Grange had done "immense" good in curbing "the abuses and wrongs" involved in the "outside business" of the farm. They had not joined the Grange "to fight manufacturers and merchants forever"; nor had they joined "to make storekeepers and manu-facturers out of the farmers." If any farmers had surplus money that they were itching to invest, they should put it into their fields and livestock. Prof-itable farms: that was the "true and sound policy." To push for much more by investing in non-farming enterprises or by taking strong actions against townsmen would destroy the Grange as a common institution for farmers (L. P. N., 1875, p. 5). Most farmers, however, did not want the Grange to be an updated version of the traditional agricultural society. Their commit-ment to the organization rested on the Grange's commitment to reforming the public policies and commercial practices of non-farmers that set the conditions in which the outside business of the farm was conducted.

One was the voice of the farmers' movement, and the other, the Grange of the future. Between the call to wage war against townsmen on all sides and the call to mind the farm, grangers were torn. Many grangers agreed on one thing though: "none but plain and practical farmers" and their interests belonged in the Grange (Several Patrons, 1875, p. 5). The Patrons Manufacturing Association would be launched over their objection. Orga-nized on the joint stock principle (like some cooperative stores) and sup-ported by a small group of wealthy individuals, it was not, strictly speaking, a mutually beneficial cooperative venture that belonged to the farmers.

From Statewide Economic Cooperation to Mutual Education

As 1875 entered its final months, the Indiana State Grange's business fea-ture was showing signs of advancing backwards. Farmers were saving money, but cohesion was fraying. There was still reason for optimism though. The "multiplicity of council" among delegates to December's annual conven-tion might enable the State Grange to arrive at an "advanced position" on the issues. The meeting promised to be the "most important" ever held in Indiana (J. N. S., 1875, p. 5).

Three results emerged from the proceedings. A new slate of officers was elected; no city farmers remained. The Patrons' Manufacturing Association received a begrudging blessing. The State Grange was "strenuously opposed" to subordinate and county granges subscribing for shares, but it could not prevent members, as private individuals, from buying stock (Indiana State Grange, 1875–1876, pp. 42–43). Those were the two easy decisions. Grangers were unable to reach agreement on what to do about the State Purchasing Agency. There were too many types of cooperative stores—stores affiliated with the county councils, joint stock county stores, and cooperative stores that were sponsored by granges within townships—each of which had a different relationship to the State Purchasing Agency. Some grangers were opposed to cooperative stores altogether. The convention was compelled to adjourn. When it reconvened in January, the Executive Committee was instructed to close the State Agency.

Responding to rumors that the State Agency might be abolished, a combination of retailers offered to create a Patrons Supply Depot (complete with a system of county agents) to furnish dry goods, household wares, and farm implements. The proposal was well calibrated for achieving compromise. As a centrally located and large scale distributor, the Patrons Supply Depot could meet the needs of any cooperative stores that wanted to use it. At the same time, since it was not an official Grange agency, there was no obligation to use it. Cooperative stores could use other suppliers as they saw fit. Grudgingly, opponents of all cooperative ventures could also accept the affiliated, but independent status of the Patrons Supply Depot. Concrete details about the plan were slow to emerge, however, so the State Grange granted conditional approval, pending a final decision by the Executive Committee (Mitchell, 1876). In the meantime, farmers' satisfaction with Tyner's work and opposition to creating more middlemen fueled discontent. The grumbling turned to howling when it became evident that the grangers had been duped.

The Patrons Supply Depot was a scheme advanced by "capitalists of other States" who had "no further interest in" the Indiana State Grange's economic cooperation "than the money they can make from it" (Seybold, 1876, p. 5). A bit of entertaining at the bars and theaters of Indianapolis had helped to convince grangers of their good intentions, but the two men responsible "were so overjoyed at the prospect of future profits that they could not help boasting" to others about their accomplishment. As word of the attempt to foist a monopoly on the State Grange circulated through the countryside, the *Indiana Farmer* "received letters from hundreds of granges, all breathing one spirit" of outrage ("The Matter Settled," 1876, p. 5). In their next breath, farmers demanded that the State Purchasing Agency be restored to its full powers. Then their outrage turned on the Executive Committee. Adhering strictly and faithfully to the State Grange's instructions, the Executive Committee continued the dismantling of the State Purchasing Agency. Authorized to make the final decision on the Patrons Supply Depot, the Executive Committee rejected it. Grangers were now deprived of a statewide institution for ordering and supply ("Meeting of Executive Committee of State Grange," 1876).

Grangers throughout the state were not pleased with that outcome. During the past year, the Executive Committee had tried repeatedly to "compel the Patrons to deal through County Agents." It had "never been satisfied" with the State Agency, despite the fact that, no matter how one figured it, maintaining supply houses and agents in each of Indiana's 92 counties could not compare to the State Purchasing Agency's efficiency (Noble, 1876, p. 5). There simply was no more cost-effective alternative to the centrally located distributor. Besides, the Executive Committee's abolition of the State Purchasing Agency begged the question: from whom, and where—at home, at Indianapolis, or abroad—were grange-affiliated county agents to order supplies (Grange, 1876, p. 5)?

Outside official channels, the call went out for another convention. Delegates from at least 20 counties assembled at Indianapolis in early February. After denouncing "the interference of the 'grip sack gentry' from the East" in the Indiana State Grange's affairs, they appealed for the State Purchasing Agency's full restoration ("The Patrons in Council," 1876, p. 5). A week later, their appointed delegation met with the Executive Committee. After a "mutual exchange of ideas" they departed, "feeling satisfied that the Executive Committee fully understand the responsibility resting upon them, and the difficulty in carrying out the action of the State Grange in relation to the business agency" (Mitchell & Jones 1876, p. 5). Although it seems unlikely that agreement could have been reached, it was. Officially, Tyner's Agency was to be closed in May; unofficially, Tyner was granted permission to carry on business, in conjunction with the now up-and-running Patrons' Manufacturing Association ("The New Business Arrangement," 1876).

The joint stock manufacturing venture needed all the publicity and order-boosting support that it could get from the well-known State Grange Agent. As late as mid-June, 1876, stock subscriptions remained unfilled. Despite enthusiastic predictions about the new Eureka mower, the only implement in production, orders did not match expectations. The leading partner of the Patrons' Manufacturing Association returned from the Centennial Exposition to face lawsuits for debt collection. The joint stock manufacturing venture was going down; it would bring the State Purchasing Agency with it ("The New Manufacturing Enterprise," 1876).

Controversy and confusion plunged Tyner's operations into disarray. Was the State Purchasing Agency open or closed? In May, the Executive Committee had instructed Tyner to close the Agency. His remaining goods were to be sold by the Patrons' Manufacturing Association at 20 percent commission. Instead of dwindling to nothing, though, Tyner's inventory kept growing. By October, with barrels of salt, woolen goods, cases of shoes, and axes passing through, Tyner claimed that it was starting "to look like old times in the bustle of business" ("State Agency," 1876, p. 5). It was nothing like old times, however. At year's end, the total purchases amounted to less than $60,000, a far cry from the previous year's $310,000 in commerce (Indiana State Grange, 1876, p. 17). In their letters denouncing the Patrons' Supply Depot scheme, grangers had revealed their confusion over the distinction between it and the Patrons' Manufacturing Association. (Both, after all, involved a centrally located distributor and involved commercial interests from other states.) Tyner's relocation to the Patrons' Manufacturing Association added to the farmers' confusion and encouraged them to make their purchases through other channels.

Indeed, it was not even clear to what extent Alpheus Tyner operated as the State Grange Agent. Acting in his (new) official capacity as a medium of communication for the county-level joint stock stores, Tyner was not supposed to handle merchandise. Inevitably, he did. Railroad carloads of bulk shipments of goods were shipped to Indianapolis, their contents parceled-out, and then distributed to their final destinations. Acting semi-officially, in his old manner, Tyner did the same for the cooperative agencies sponsored by county councils (which refused to disband) and the cooperative stores run by neighborhood granges. Yet again, Tyner took orders and handled merchandise as an agent for the Patrons' Manufacturing Association under the *ad hoc* 20 percent commission agreement. Acting in various stages of official capacity, Tyner was supposed to sell the goods at different rates to the different kinds of grange-affiliated cooperative stores, individual grangers, and non-grange-joining farmers. Little wonder, that the Executive Committee "found complications complicated" when it inspected the account books prior to the 1876 State Grange convention. Which business belonged to the State Grange and which to the Patrons' Manufacturing Association? While the Executive Committee worked with Tyner to arrange the accounts "in an intelligible manner," the State Purchasing Agency was ordered closed and its inventory posted for public auction (Indiana State Grange, 1876, pp. 26–27).

The State Grange committee that reviewed the Executive Committee's actions in closing the State Purchasing Agency, delivered, at best, a mild rebuke. As the State Grange's Worthy Master observed to his audience, too many substantial grangers now agreed that farmers had "so many varied interests" stemming from geographic location "that no general system" could be sustained (Indiana State Grange, 1876, p. 6). Another year's experience with locally based cooperative stores had persuaded others that it was better to keep farmers' dollars in their home communities. Still others agreed with C. C. Post, the State Secretary. If farmers had worked to gain "a fuller knowledge of the laws of business," they would have been better equipped "to wrestle with the great questions of co-operation and transportation which have and will continue to agitate the public mind for years" (Post, 1876, p. 5). Many grangers, of course, were inclined to the opinion that farmers did not have—and never had—any business engaging in any pursuits other than farming. Regardless of the exact reasons, the State Grange was in solid agreement. The State Purchasing Agency would not get another lease on life.

Farmers had joined the Grange to save money. With the deliberate disruption and dismantling of the State Purchasing Agency, the Grange failed to deliver ("Importance of the Order," 1877). Within months it was "painfully evident that in many localities the interest and zeal formerly manifested in the Grange" was no more (Cory, 1877, p. 7). Enrollment plummeted by two-thirds over the next two years, and continued to wane with each passing year. Repeated appeals were made for reinstating the State Purchasing Agency or for creating some other institution to carry out economic cooperation on a statewide basis. All appeals were denied without, apparently, much consternation among the farmers who made up the State Grange (Indiana State Grange, 1878, p. 41). Among most of the farmers who remained in the Grange, regard for their personal self-interest and the well-being of nearby towns trumped their concern for the independence of other farmers. If there were farmers who had yet to harbor this suspicion, the refurbished style of grange-sponsored cooperation in local communities would confirm it.

Much of the State Grange leadership's difficulty with the State Purchasing Agency (and economic cooperation in general) was its inability to discipline the rank and file membership. Farmers paid little regard to directives and interpreted Grange doctrine as it suited their particular purposes and circumstances. A good portion of the farmers' disorderliness stemmed directly from the workings of the county councils. Unauthorized by the National Grange, the county councils had been inspired by the Illinois State Farmers' Association. They created a host of problems for the Grange. Fundamentally, they were too democratic, too receptive to the narrowly conceived interests and short-sighted opinions of common farmers. The county councils were also too open. Grangers' secret business arrangements were debated publicly, stirring up controversy between grangers and unaffiliated farmers, and among rival tradesmen of the towns. The county councils' openness also made them potent. Non-farmers were permitted inside the gate. These were the men who led the way in organizing grange-sponsored cooperative stores, stockyards, and other upstart competitors to established town-dwelling interests. The county councils, in sum, played a large role in carrying the farmers' movement into the Patrons of Husbandry and in inflaming the animosity between farmers and townsmen.

The Pomona grange was intended to put the county councils out of business. On its face, the Pomona grange was simply a sanctioned replacement at the county level. Few people thought that a major change was underway when the State Grange authorized its creation at the annual meeting of 1874 (Indiana State Grange, 1874). Most people thought it was a means of strengthening the business feature by conferring legal recognition upon county-level operations. They expected the Pomona grange simply to "take the place and do the offices of the County Council, under a new name" without making any "doubtful or unsatisfactory change in the method" of conducting business affairs. As with the State Purchasing Agency, though, the devil was in the details of implementation ("The Pomona Grange," 1875b, p. 5).

The same season that the Executive Committee tried to reinterpret the goods and grocery-ordering function out of the State Purchasing Agency, it designed the rules for the Pomona grange. Its actions went well beyond legalizing the existing mode of operation used in the county councils. The democracy and openness of the county councils were purged. Eligibility to membership in the Pomona grange was restricted to grangers who had advanced to the highest rank in the neighborhood grange. As a condition of membership, they were required to pay a three-dollar initiation fee and double the annual dues. Already grumbling about fees and dues, farmers were unlikely to voluntarily tax themselves to support representation in an objectionable institution. Men (and their wives) who sought to join the Pomona grange would go on their own dollar. Odds were they would be wealthier farmers who belonged to granges near the county seat. As few as nine men and four women could establish a Pomona grange (one per county). There was no stipulation for proportional representation from a county's subordinate granges ("Pomona Grange," 1875a).

As details about the Pomona grange's exact design emerged in the summer of 1875, some grangers demanded its suspension. The most damning and revealing charge was that it was an attempt to create an "aristocracy" within the organization, even though the prevailing opinion among farmers favored an "equal footing" for all (McGaughey, 1875, p. 5). Opponents of the Pomona grange hailed the County Council as "the most popular, the most democratic and purely representative body" known to farmers (Collier, 1875). Other farmers rejected the notion that the Pomona grange represented an improvement in how business affairs would be handled. The county councils were better-suited, since men could be chosen with an eye to their "business qualifications," as distinct and separate from the other desirable qualities and talents involved in leadership of a local grange club (Sawdon, 1875, p. 5). By the close of 1875, Pomona granges were established in only 18 of Indiana's 92 counties. County councils refused to disband and farmers continued to support them (Indiana State Grange, 1875–1876, p. 8).

Savings bolstered Hoosier farmers' attachment to the county councils. The goods and groceries of the State Purchasing Agency were the councils' lifeline. The Executive Committee's assault on the State Purchasing Agency in early 1875, therefore, had been an indirect strike at the county councils. Alerted to the danger, the county councils pushed for, and obtained over against the Executive Committee's preference, mechanisms to establish closer relations with the State Purchasing Agency. The two-stage order-pooling and county agent/purchasing agency system about which some grangers had grumbled had solidified the county councils' role in the business feature. Their representatives, the State Council of Purchasing Agents, worked with Alpheus Tyner to reach agreements with wholesalers, and their selected agents handled order-taking and distribution in the counties. Under the arrangement, farmers dealt through a county-level middleman, but they received greater savings than was possible through local merchants. It was significant—although many grangers seemed not to fully realize just how much—that it was their middleman, an agent who was part of a broader system that was accountable to the rank and file of the Grange.

Even at the peak of their cooperation with Tyner's State Agency in 1875, the county agencies were feeling their own way. Farmers did not think it necessary to buy exclusively through the State Agency. Proximity to major cities and shipping routes made a great difference in the savings derived by grangers in different regions of Indiana. Tyner could not always offer the best deal. True to the basic principle of economics, farmers went for the lowest price available on a cash-only basis. Concerted efforts by town merchants and their suppliers to sell goods "cheap as dirt" gradually achieved the desired effect (J. H. W., 1877, p. 7). Despite cautionary warnings that the price reductions would be temporary unless a grange system was sustained "as a check upon merchants and speculation," grangers continued to engage in outside purchasing (McCoy, 1876, p. 5). The Executive Committee, of course, encouraged the tendency by insisting that local granges try the nearest merchant for their groceries and goods. In the aftermath of the Supply Depot scheme, grangers had little else remaining as an option.

The State Purchasing Agency's lingering death through 1876 put the county councils out of business. Political rancor from the Greenback movement played its part, but the crux of the matter was that, without the State Purchasing Agency, the county councils had limited leverage on local merchants (Caldwell, 1875; Gates, 1875). Pomona granges began replacing the councils as the county-level grange institution. On the whole, they were open to the idea of limited cooperation (to correct the worst abuses or to obtain modest savings on selected items), but they opposed the kind of all-out war waged by the county councils against established local merchants. As a consequence, county-level grange agencies and cooperative stores that pooled farmers' purchasing power were abolished or reorganized as joint stock stores.

Local cooperation furnished "all that can be reasonably expected in the way of a reduction of prices" (Indiana State Grange, 1877, p. 9). That was the message the State Grange's Worthy Master gave to grangers who demanded the State Purchasing Agency's restoration. The cooperative plan he had in mind did not use Grange dues to give farmers' stores a competitive advantage over town merchants. It did not even offer goods at reduced rates, beyond what cash payment secured. Instead, it required grangers to pay the "usual prices for goods at the time of their purchase." At the end of the year, the store's profits would be divided among its subscribers "in proportion to their purchases" (Shankland, 1876, p. 5). The new plan was the kind of economic cooperation that received the sanction of the National Grange ("Co-operative Associations as Savings Banks," 1876). It was not, however, the kind of economic cooperation that had brought Indiana's farmers into the Grange.

Nevertheless, the new form of cooperation spread. In it, traveling Grange lecturers found support for the claim that "the success of our Order depends upon the success of the business feature." Granges with "energy and zeal," they reported, "almost invariably [had] a co-operative store" ("Grange Lecturers—The Business Feature," 1878, p. 7). The store was not the reason; it offered neither savings nor profits enough. Most folded within a year or two. Farmers who were not grangers refused to patronize the cooperative stores (J. A. P., 1877, p. 7). Who could blame the farmers? The grangers sold them out by tearing down the State Purchasing Agency and eliminating the county councils. And, there was little difference between granger stores' prices and the prices charged by town merchants. The town merchants, at least, were willing to extend a line of credit to struggling farm families.

Ultimately, the cooperative stores' failure did not matter. The Grange made it clear that it would not carry the farmers' movement and common farmers were abandoning it. Grangers made, at best, limited efforts to retain them. What they wanted in their ranks was "the right kind of material": "substantial farmers who have come to stay" (Almond, 1877, p. 7). That kind of farmer would have agreed with *Indiana Farmer* editor J. G. Kingsbury, when, in 1878, he told them that the Grange was not "fulfilling its highest mission in starting stores" (Kingsbury, 1878, p. 7).

That kind of farmer would have agreed with the State Grange's Worthy Master who, in the early 1880s, pointed to the spread of "mutual benefit" or "cooperative" insurance against death and fire as evidence that the "business feature" had succeeded in the Grange (Indiana State Grange, 1881, p. 12). He also would have shared in the Worthy Master's gratitude that the "excitement and furor" over "questions of state and finance" during the Grange's early years had given way to an "advanced position." The new granger would hail the "high and noble purposes" of a properly conducted grange: the grange as "not only a good agricultural society," but, also as "a lyceum, a school, where the latent powers of the mind are developed," as a "common center from which shall emanate influences that will elevate the social, intellectual, and moral growth of society, and from which should flow a public spirit that shall advance the general welfare, harmonize discordant elements, and thus hasten the day when good-will and fraternal regard shall characterize all" (Indiana State Grange, 1882, pp. 12–13).

It was a noble sentiment, even if relatively few farmers remained in the Grange to hear it. It was also typical of the rhetoric that floated in the air at the variety of voluntary society meetings attended by town-dwellers. They maintained a studied attitude of indifference toward the plight of nonmembers and their fellow townsmen. If at all possible, they had even less regard for the common farm family of the countryside. Fully respectable and fully in accord with the civic temper, the grangers had gone to town without leaving the farm.

Reflections on Civic Learning

In the post-Civil War era, Indiana's farmers were struggling to adapt to sudden changes in their economic environment. High staple crop prices existed no longer. Saving money on commercial transactions became all the more crucial to maintain their independence. Yet, high prices were imposed by commercial networks that stretched across the nation. Encountered by farmers through transactions with local agents, middlemen, and merchants, the networked organization of regional industries disrupted the traditional ideal of a free market economy: that buyers and sellers were free agents who acted independently when bargaining. Constrained and empowered by combinations, merchants and middlemen were not free agents; farmers were. As isolated individuals, they had no choice but to submit to the terms and conditions offered by those with whom they transacted business.

For mutual protection farmers needed bargaining leverage from an interest-based organization. From a civic perspective, that kind of farmers' organization could not be permitted. The organization of networks for credit, purchase, and supply among non-farming industries was approved. They gave commercial transactions a degree of reliability across broad geographic spans. Equally important, numerous networks existed within and across particular industries, with competition sufficiently robust (in the 1870s, at least) to yield astonishing rates of innovation. As an industry, farming was different. Farmers constituted the majority of the population—the vast majority in many communities—and most industries depended heavily on agriculture in one way or another. There were too many farmers, with too much economic clout for them to be combined effectively into a single organization. Had the farmers' economic cooperation reached its fullest potential in local communities, it could have impaired seriously economic progress. Or, and contrary to intentions, if farmers' economic cooperation had succeeded on a regional scope, it could have accelerated the concentration of economic power into fewer hands—and places—by giving particular networks the decisive advantage over others.

The late nineteenth century United States is famous for its rapid industrialization and its small town life. Both were being made possible by exploiting farmers and farmers knew it. Determined to maintain their own independence on the farm, they rallied around economic cooperation. As a pocketbook matter, the farmers' solution made perfect sense. They saved money and injected competition into the local scene. In doing so, however, they chose to lend their mite to the economic fortunes of New York City, Chicago, St. Louis, and Indianapolis. Small town merchants and manufacturers could not compete against big city wholesalers and national corporations. Understandably, leading townsmen opposed farmers' economic cooperation. It was a genuine threat to their towns' livelihood, one sufficient to justify applying peer pressure on other townsmen to convince them that their true self-interest lay in closing ranks against the Grange. Civic allegiance demanded action on behalf of one's town.

For farmers' economic cooperation to succeed, institutional mechanisms and singularity of purpose were needed. Between the State Purchasing Agency and the county councils, the Indiana State Grange had a hub-and-spoke structure that worked. As long as farmers stood behind it, there was little that town merchants and middlemen could do to counteract it. One thing they could do, however, was to cut rates and prices. From that point on, the failure of economic cooperation lies with the farmers. A voluntary organization made up of voluntary organizations, the Indiana State Grange depended almost entirely on each member's commitment to the well-being of other farmers. Most farmers were too independent-minded, too heedful of self-interest (or too close to the margin of subsistence), and too local in outlook to sustain a statewide campaign of economic cooperation. Farmers had the machinery, but lacked the will.

Economic cooperation taught Indiana's farmers that they did not possess a single overarching economic interest as farmers. Marching to picnic rallies it was easy to imagine that they shared common cause. The common cause was real; applying it to particulars was the problem. Every farmer was a resident of some place, and that place had interests that conflicted with other places, particularly Indianapolis. Some farmers lived close to central towns; others lived in the remote countryside. To varying degrees, farmers earned income from supplying local towns or almost entirely from distant markets. Most farmers were impoverished smallholders who wanted tools, household goods, clothing, and groceries for their families. Wealthier farmers had those things; they wanted to purchase labor-saving machinery. Some farmers, moreover, were not strictly farmers; practicing other pursuits on the side (or vice versa) they had interests that interfered with farmers' solidarity. Farmers were unable to reconcile their conflicting interests as members of a common organization. Ultimately, the pull of civic loyalties upon leading farmers played the decisive role in economic cooperation's demise.

The dismantling of grangers' economic cooperation was a blow to a triple set of traditional American ideals: democracy, rough economic equality, and the farmer's independence. It was a victory for another triple set: republicanism, progress, and the interdependence of town and country. It was also a victory, of sorts, for the Grange. Voluntary associations originated in towns. As an association for farmers, the Grange was intended to bring them into the currents of the town-dwelling life. By excluding economic cooperation, and embracing social and educational features as its primary reasons for existence, the Grange endured to accomplish that mission. The Indiana State Grange's Worthy Master was correct to liken the Grange of the 1880s to an agricultural society and a school. Neither institution engaged economic and political controversies; both advocated a brand of conservatism that blended traditional morality and economic progress. Those were the ideal characteristics of educational institutions, voluntary associations, and the town-dwelling middle class in the second half of the nineteenth century. For Indiana's new grangers, an institution that sought to cultivate the hearts and minds of the best citizens of the countryside should aspire to nothing less.

References

A. L. D. (1875). The executive committee. *Indiana Farmer, 10*(24), 5.

Almond, C. (1877). [Untitled letter to the editor]. *Indiana Farmer, 12*(21), 7.

Ayers, L. S. (1874). Grangers! [Advertisement]. *Indiana Farmer, 9*(26), 4.

Billingsley, J. J. W. (1875). A great loss. *Indiana Farmer, 10*(9), 5.

Business of the executive committee of P. of H. (1875). *Indiana Farmer, 10*(16), 5.

Caldwell, J. H. (1875). Boone County. *Indiana Farmer, 10*(47), 5.

Cerny, G. (1963). Cooperation in the Midwest in the granger era, 1869–1875. *Agricultural History, 37*(4), 187–205.

Clay, J. C. (1876). Harrodsburg Grange. *Indiana Farmer, 11*(43), 5.

Collier, R. (1875). Pomona granges. *Indiana Farmer, 10*(29), 5.

Comstock, H. (1875). Trial of plows in Wabash Co. *Indiana Farmer, 10*(18), 7.

Co-operative associations as savings banks. (1876). *Indiana Farmer, 11*(45), 5.

Cory, R. F. (1877). Resolution of Spring Valley grange. *Indiana Farmer, 12*(11), 7.

Doty, D. (1873). Club in Pike County. *North Western Farmer, 8*(5), 109.

Fisher, A. B. (1875). Some things inexpedient. *Indiana Farmer, 10*(23), 5.

Forward. (1875). The state agent. *Indiana Farmer, 10*(2), 5.

Futile opposition. (1876). *Indiana Farmer, 11*(40), 5.

Galbraith, M. V. (1873). Objects of the order. *North Western Farmer, 8*(3), 64.

Gates, D. W. (1875). [Untitled letter to the editor]. *Indiana Farmer, 10*(44), 5.

The Grange. (1874). *Indiana Farmer, 9*(18), 4.

The Grange. (1877). *Indiana Farmer, 12*(23), 7.

Grange. (1876). The state agency. *Indiana Farmer, 11*(2), 5.

The Grange at the state fair. (1875). *Indiana Farmer, 10*(41), 5.

Grange lecturers—the business feature, etc. (1878). *Indiana Farmer, 13*(25), 7.

The Grange organization and what it has done for the farmer. (1874). *Indiana Farmer, 9*(34), 4.

Granger. (1875). [Untitled letter to the editor]. *Indiana Farmer, 10*(3) 5.

A granger's dream. (23 April 1874). *Dearborn Independent.*

Grant, G. W. (1875). [Untitled letter to the editor]. *Indiana Farmer, 10*(7), 5.

H. (1875). The plow question. *Indiana Farmer, 10*(36), 5.

Harvey, A. C. (1873). [Untitled letter to the editor]. *North Western Farmer, 8*(7), 160.

Howard County grange association. (1876). *Indiana Farmer, 10*(13), 5.

Importance of the order throughout the state. (1877). *Indiana Farmer, 12*(9), 7.

Indiana State Grange. (1874). *Fourth Annual Proceedings.* Indianapolis, IN: Author.

Indiana State Grange. (1875–1876). *Fifth Annual Proceedings.* Indianapolis, IN: Author.

Indiana State Grange. (1876). *Sixth Annual Proceedings.* Indianapolis, IN: Author.

Indiana State Grange. (1877). *Seventh Annual Proceedings.* Indianapolis, IN: Author.

Indiana State Grange. (1878). *Eighth Annual Proceedings.* Indianapolis, IN: Author.

Indiana State Grange. (1881). *Eleventh Annual Proceedings.* Indianapolis, IN: Author.

Indiana State Grange. (1882). *Twelfth Annual Proceedings.* Indianapolis, IN: Author.

Jackson, J. F. (1875). [Untitled letter to the editor]. *Indiana Farmer, 10*(2) 5.

J. A. P. (1877). Co-operation. *Indiana Farmer, 12*(4), 7.

J. C. A. (1876). What of the night? *Indiana Farmer, 11*(4), 4.

J. H. W. (1877). Co-operation in Decatur County, etc. *Indiana Farmer, 12*(6), 7.

J. K. (1875). [Untitled letter to the editor]. *Indiana Farmer, 10*(4), 5.

J. S. D. (1875). The manufacturing association. *Indiana Farmer, 10*(45), 5.

J. N. S. (1875). Coming State Grange. *Indiana Farmer, 10*(49), 5.

Kingsbury, J. G. (1878). The future of the Grange. *Indiana Farmer, 13*(1), 7.

L. P. N. (1875). Opposes the movement. *Indiana Farmer, 10*(48), 5.

The matter settled. (1876). *Indiana Farmer, 11*(7), 5.

McCoy, A. H. (1876). Stand by your Grange. *Indiana Farmer, 11*(32), 5.

McGaughey, J. E. (1875). Pomona granges. *Indiana Farmer, 10*(34), 5.

Meeting of the executive committee of the State Grange. (1873). *Indiana Farmer, 8*(5), 6.

Meeting of executive committee of State Grange. (1876). *Indiana Farmer, 11*(6), 5.

Miller, J. R. (1875). Come to stay. *Indiana Farmer, 10*(9), 5.

Mitchell, R. (1875). The state agency. *Indiana Farmer, 10*(3), 5.

Mitchell, R. (1876). Proceedings of executive committee. *Indiana Farmer, 11*(4), 5.

Mitchell, R., & Jones, A. (1876). Proceedings of the executive committee. *Indiana Farmer, 11*(7), 5.

The new business arrangement. (1876). *Indiana Farmer, 11*(21), 5.

The new manufacturing enterprise. (1876). *Indiana Farmer, 11*(22), 8.

The new method for business. (1875). *Indiana Farmer, 10*(5), 5.

Newsom, J. Q. A., & Mitchell, R. (1874). Proceedings of the executive committee. *Indiana Farmer, 9*(52), 4.

Newsom, J. Q. A., & Mitchell, R. (1875). Proceedings of the state executive committee, P. of H. *Indiana Farmer, 10*(22), 5.

Noble, W. I. (1876). Attempted monopoly—the meeting of February 7th. *Indiana Farmer, 11*(5), 5.

Nordin, D. S. (1974). *Rich harvest: A history of the Grange, 1867–1900.* Jackson: University Press of Mississippi.

The Oliver chilled plow. (1874). *Indiana Farmer, 9*(39), 4.

The opportunity offered. (1875). *Indiana Farmer, 10*(43), 4.

Patron. (1875). Let all be faithful. *Indiana Farmer, 10*(21), 5.

A Patron. (1875). [Untitled letter to the editor]. *Indiana Farmer, 10*(45), 5.

The Patrons in council. (1876). *Indiana Farmer, 11*(6), 5.

Patrons of Husbandry. (1875). *Indiana Farmer, 10*(14), 5.

Pomona grange. (1875a). *Indiana Farmer, 10*(18), 2.

The Pomona grange. (1875b). *Indiana Farmer, 10*(37), 5.

Post, C. C. (1876). Cultivate the social and intellectual faculties. *Indiana Farmer, 11*(17), 5.

Proceedings of the executive committee of the State Grange, held at Indianapolis, Jan. 26th to 29th. (1875). *Indiana Farmer, 10*(5), 4.

Proceedings of the state council of county purchasing agents of P. of H. (1875). *Indiana Farmer, 10*(9), 4.

R. H. K. (1873). Opposition to the Grange. *North Western Farmer, 8*(7), 160.

Ross, T. (1876). [Untitled letter to the editor]. *Indiana Farmer, 11*(15), 5.

R. S. (1875). Oliver chilled plow. *Indiana Farmer, 10*(24), 5.

Sawdon, G. W. (1875). Pomona grange. *Indiana Farmer, 10*(29), 5.

S. B. B. (1875). Stand by your principles. *Indiana Farmer, 10*(30), 5.

Several patrons. (1875). The next state secretary. *Indiana Farmer, 10*(46), 5.

Seybold, J. L. (1876). The attempted monopoly. *Indiana Farmer, 11*(5), 5.

S. G. (1875). The manufacturing association. *Indiana Farmer, 10*(48), 5.

Shankland, R. R. (1876). Benefits of co-operation. *Indiana Farmer, 11*(43), 5.

Shiloh. (1875). Wabash County. *Indiana Farmer, 10*(46), 5.

South Bend Iron Works. (1874). Important to farmers. [Advertising supplement]. *Indiana Farmer, 9*(52), 2.

State agency. (1876). *Indiana Farmer, 11*(26), 5.

State agents. (1875). *Indiana Farmer, 10*(32), 5.

State purchasing agency. (1874). *Indiana Farmer, 9*(24) 4.

Thomas, V. (1873). Union organization of Porter Co. *North Western Farmer, 8*(3), 65.

Thornbrough, E. L. (1965). *Indiana in the Civil War era.* Indianapolis, IN: Indiana Historical Bureau and Indiana History Society.

Thorntown. (1873). Boone County. *Indiana Farmer, 8*(2), 5.

Watchman. (1875). The coming State Grange. *Indiana Farmer, 10*(47), 5.

What has been done. (1876). *Indiana Farmer, 11*(20), 5.

Why don't somebody howl? (1874). *Indiana Farmer, 9*(36), 4.

7

Bringing Town and Country Together for Progress at the County Fair

The Grangers and Progress

Grangers pledged to work to put American society on sound principles and they meant it. They viewed themselves as participants in true progress. On this account, all genuine and permanent social improvement began with each person's commitment to self-improvement and determination to act in accordance with his best long-term interest. From this starting point, all aspects of society—material and moral, social and intellectual, personal and public—would improve steadily, incrementally if need be, but together and forward. It was an undirected, but rational vision of social advancement and a time-honored doctrine in the American tradition. By subverting the farmers' movement, the grangers signaled that they knew what was expected of good citizens. Instead of taking aggressive action on behalf of a broad program of political and economic reform that might jeopardize the economic interests of towns, Grangers would improve society by first improving themselves.

It was here that grangers parted ways with the farmers' movement. Even if farmers triumphed in public policy or economic cooperation, lasting re-

Civic Learning through Agricultural Improvement, pages 151–176
Copyright © 2011 by Information Age Publishing
151

form could not be achieved. How could it be until farmers changed their ways? Farmers had abandoned the principles of "rigid economy and self-denial" that had guided them for generations. Lulled by high staple crop prices and the "superabundance of cheap money" during the Civil War, they bought more land and new machinery (Neidhart, 1875, p. 6). While the boom times lasted, they made useless expenditures on the household and for fine clothes to wear on trips into town. Worse still, they failed to change their approach to farm management. When staple crop prices sank, farmers persisted in raising the same staple crops; then, to get by through the year, they borrowed at high rates of interest in the hope that the next harvest would be a good one and that high prices would return. Farmers, in short, were guilty of the luxury and corruption that they decried in town-dwelling merchants and middlemen (W. H. S., 1874).

On grangers' account, farmers could claim righteous indignation at being victimized, but they could make little claim to virtue in their business practices. Were not most farmers gambling on precarious world markets, hoping for that occasional bumper season that also brought large returns on single cash crops (such as wheat, corn, and hogs)? Were not they pursuing easy money by planting crops that required comparatively less work, while neglecting more labor intensive alternatives? Were not many farmers skinning the land with the intent of moving on after a few years, having taken much, but returned little to the community? Did not everyone know of more lucrative, stable, and socially useful ways of making the farm pay? Through the previous generation, the unrelenting message of agricultural papers and societies had been that combinations of crop- and livestock-raising could provide stable income, replenish the soil, and supply the needs of local town-dwellers' tables and industries. What farmers had not heard, time and again since boyhood, that the sure road to wealth was the long one, the one that called for growing a farm through hard work, self-study, and pay-as-you-go improvements? These were tough questions to put to struggling farmers.

As the farmers' movement faded, the number of men willing to point fingers at themselves or, at least at other farmers, grew. In the meantime, demands for economic justice were countered at grange meetings through instruction in agricultural and political economy. Living within one's means by purchasing with cash only was upheld as one of the highest virtues ("Don't Go In Debt," 1874). It was followed closely by the injunction to get "the largest return for the smallest outlay" on every acre of farmland (Hazlet, 1874, p. 4). Personal reformation—"going back to first and sound principles"—in all aspects of one's life and business was insisted upon. Avoiding "the extravagance of ideas as well as living" went with it (Old School,

1873, p. 11). Frugality and simplicity, careful attention to business, and self-improvement in accord with propriety: these elements comprised the core of grangers' prescription for adjusting to the modern industrial economy. It promised a sure and steady income (although not lucrative profits), one that correlated with the diligence and intelligent management that each man put into his farm. Each man was entitled to receive a just return on his toil and investment, no more and no less. That was the true measure of personal economy, and it applied to all businesses, farming included.

Toward the town's industries, grangers and like-minded farmers maintained the same attitude. Like their land-skinning countrymen, too many townsmen pursued gain too avidly, without heeding its broader consequences, especially townsmen of influence and public standing. The economic depression of the 1870s, all farmers knew, had its roots in the mania for railroad building and manufacturing investment that had possessed townsmen in the wake of the Civil War. The county fair was an annual reminder of how readily greed trumped public morality. Sponsored by the most influential townsmen and given over to horse-racing, gambling, drinking, and sideshows featuring every kind of depravity, the county fair was no place for respectable women and children. It was no place for respectable farmers either. Credentials as a scientific agriculturist were not needed to recognize that "snakes and serpent charmers, fat women and dwarfs, dancing bears and lazy monkeys" did little to develop "new ideas in agriculture" ("The Hendricks County Fair," 1868, pp. 222–223). Little wonder, then, farm families' attendance diminished as county fairs grew in scope and scale.

The county fair, of course, was more than the largest civic event of the year: it was the principal economic event of the year. The thousands who gathered for the horse races and revelry of a thriving fair served as a powerful stimulant to the host town's economy. The show itself—the displays of livestock, home industries, mineral wealth, and forest products—contributed to future economic development by making known to the world a county's advantages for manufacturing, investment, and relocation. Town boosters loved to proclaim that there were no "better data from which to approximate at the thrift and enterprise of a locality, than the interest taken by the people in the yearly exhibition of the products of the soil and the workshop" ("Porter County Report," 1874, p. 200). Almost as frequently, they complained that there was "no good reason" why their county—given its bounteous land, its excellent location, its rail and water facilities, and its towns—could not boast "as good a county fair as any in the State" ("Cass County Report," 1874, p. 161). Town boosters were determined to bring the loom and the anvil into proximity with the plow as quickly as possible; as fair managers they were willing to sanction a little popular vice for the greater good of enhancing their town's prospects for manufacturing and economic development.

Respectable farmers' appreciation for home manufacturing grew with each passing year, but the degree to which agricultural societies were willing to sacrifice public morality and useful improvement to advance it was too much. The county fair was supposed to honor the contributions all forms of productive labor made to the common prosperity. Symbolically, it was a tangible expression of the ideal of a harmony of interests, an annual ritual that displayed the interdependence of the respective industries of town and country. Practically, the county fair was supposed to function as an educational agency, one that inspired people to self-improvement in their callings by exposing them to emerging knowledge and technological innovations. What happened at the town-boosting fairs of the 1870s bore little direct relation to productive labor, let alone agricultural improvement. Grangers had shown their civic spirit by thwarting the farmers' movement. Now they insisted that leading townsmen show theirs. It was the last institution to feel the power of Grange reform, but the county fair furnished the most visible testimony of how town and country came together for progress in late nineteenth-century Indiana.

Fair Discontent Among the Farmers

As members of the agricultural societies, leading farmers had gone along with the decision to open county fairs to horse racing and sideshows. At best, it was a grudging concession to financial necessity. Prominent farmers had little love for fast horses, less for the town-dwelling masses, and only disdain for the sideshow tents and gambling stalls. They were willing to tolerate a bit of these things for the sake of supporting agricultural improvement. Some, no doubt, thought that they would have to put up with one, at most two, degenerated fairs to pay for the fair's land and facilities. Others were unprepared for the scale and degree of degeneracy that filled the revived fairs. Farmers who had fairly accurate suspicions about what would transpire were simply outvoted. In the immediate aftermath of the Civil War, the town-boosting imperative trumped all other considerations.

In 1868, a double-barreled blast in the *North Western Farmer* (predecessor to the *Indiana Farmer*) kicked off the clamor against the depravity and excesses of the town-boosting fair. Two separate articles sounded the refrain that would be carried for the next 10-15 years (and more in some places): "OUR AGRICULTURAL FAIRS, ARE NOT AGRICULTURAL FAIRS." As with Grange reform generally, critics of the county fairs called for a return to first principles. Fair managers could start by asking the simple question:

"What is the purpose of an agricultural fair" ("The Hendricks County Fair," 1868, pp. 222–223)?

The preferred answer was more generous to town-dwelling interests than one might suspect. No one called for a fair devoted exclusively to agricultural matters. Everyone accepted that the agricultural fair had a broader mission to advance the people's knowledge and use of "those things that go to promote the general comfort and welfare of mankind" (Stevenson, 1868, p. 276). To accomplish that aim, an agricultural fair was supposed "to instruct and entertain the people" in ways that did not overstep "the boundaries of strict moral propriety." In other words, an agricultural fair should encompass more than the strictly agricultural in its displays and functions, but no matter what else it did, it had to promote only the useful and the good ("The Duty of Fair Managers," 1870, p. 7).

Practicality and morality, the useful and the good, was there any difference? For denouncing how their opposites demoralized the agricultural fair, no distinctions were necessary. What could be more useless than spending money to wager on a horse race? What could be more depraved than the sight of hundreds of drunken fools clamoring for the privilege? What could be more degrading than having to walk through them to get to Floral Hall? Responding to "the yells of side-show criers" and stepping into a tent might have supplied it. For a nickel, one could be "disgusted with hideous, distorted pictures of hunchbacked, flat-headed, snaggle-toothed, half-naked foreigners, or some loathsome reptile coiling itself around the waist and arms of a woman." The activities in other tents were unmentionable in polite society. No utility and much harm to morality, this was what the county fair had to offer ("The Hendricks County Fair," 1868, pp. 222–223).

The solution was obvious. Agricultural societies were democratic institutions, open to anyone who paid the annual one-dollar fee (or, if reorganized, to anyone who purchased a share). All members were entitled to participate in the proceedings, to vote on the officers and major decisions. Although they were members, farmers failed to attend the meetings. By default, they left it to "the horse men to select the officers and mark out the yearly programme." Was it any wonder, then, that fast horses, saloons, and "rowdy, boisterous persons" filled the agricultural fairs? Unless and until a large number of farmers attended the meetings, they could hardly expect county fairs to reflect the real interests of agriculture (Voice from Hendricks, 1868, p. 277). Once they did attend, they could insist on useful improvements such as stock sales and practical lectures, and the elimination of everything that had an "immoral tendency" (Deal, 1873, p. 33). Farmers had the strength of numbers; if mobilized, they could "inaugurate a new system of county fairs" (Jackson, 1872, p. 164).

Hendricks County was among the first to experience the effects of organized farmers. At the start of 1869, they introduced "an innovation" into the plan of operations for the next fair: "the utter exclusion of time rings and all manner of traveling shows from the grounds." The idea met a "factious and determined opposition" from the town's merchants and businessmen. From January to September, the debate raged as to "whether horse-racing, fat women, big snakes and gaming tables aided in the advancement of agriculture." Advocates of reform were certain that they did not. The restyled "farmers' and mechanics' fair" of Hendricks County proved the soundness of their convictions with a 70 percent increase in articles exhibited. When, a few years later, Putnam County adopted the same policy, their "agricultural hall was filled for the first time in the history of the Fair" ("Hendricks County Report," 1869, p. 360; "Putnam County Report," 1873, p. 194).

Few agricultural societies followed suit. Not enough farmers attended meetings to restore the fairs to first principles. A number of agricultural societies, however, did make one concession to farmers' objections: They took "a firm, decided, and unanimous stand" in favor of protecting the "the innocent and unsuspecting" from being fleeced by "improper characters and gamesters." The hucksters were evicted from the fairgrounds ("Decatur County Report," 1873, p. 149). Generally, though, tent villages were permitted to form outside the gates. Fair-goers still had to pass through the hucksters and the agricultural societies still obtained the licensing fees that were assessed by the county. Evicting the chuck-a-luck tables and gambling stalls required little sacrifice in the name of reform.

Abolishing horse racing was another matter. Because it brought the crowds, horse racing was almost always counted as something that encouraged "the real agricultural, horticultural, and mechanical interests of the county." In Gibson County, a pledge by the agricultural society to ignore, "as far as practicable, what was simply for show and of no real utility" kept the horses circling the track in harness and under saddle ("Gibson County Report," 1871, pp. 278–279). Under greater pressure from the farming element, the Laporte County Agricultural Society divided its fair down the middle, allocating three days to practical displays only and hosting "trials of speed" on the other three days ("LaPorte County Report," 1873, p. 172). Jennings County tried the same experiment. One year of separating the horse racing from the agricultural displays was enough; the next year found the events combined, with the horses back on the track ("Jennings County Report," 1874, p. 179).

The Jennings County experience proved to be the rule. County fairs could not—or would not—go without horse racing. Instead of banning it, they tried reducing the number of races and enforcing prohibitions against gambling. Together with a ban on alcoholic drinks and side shows inside the grounds, it was a reasonable solution, an effort to find the "Golden Means" between the conflicting demands of different elements of the rural community ("Huntington County Report," 1873, p. 161). Neither the horse lovers nor the farmers were ready for compromise. If agricultural societies offered premiums for "fast-going" they were denounced by farmers; if no premiums were offered for "speed" they were denounced by the masses who wanted to see racing. Efforts "to steer between by giving small premiums"—thus discouraging professional horsemen and gamblers—"drew the abuse of the whole community" ("Vigo County Report," 1869, p. 403). With leading townsmen favoring fast horses and a growing number of farmers demanding strict morality, few people probably imagined that the county fair controversy could worsen.

When the economy collapsed in 1873, it did. Desperate to generate revenue, town merchants were unwilling to give up any of the economic activity created by big fairs. Confronted with farmers' complaints, fair managers pleaded necessity. Personally, they were opposed to "immoral devices" but "a *partially depraved* public sentiment" demanded such things. Without them, the fair would fail. Given the hard times, the horse racing ("and the fat woman, and the big bear, with an occasional wheel of fortune, swig, rope-walker, or Negro dance") was a "necessary evil" that had to be tolerated (R., 1876, p. 3). After all, did not the crowds and their dollars pay for the fair's promotion of agricultural improvement? Farmers, therefore, were well advised to leave "the pleasures of speculating and moralizing upon the influence of horse racing upon the people" to professors of moral philosophy ("Jasper County Report," 1879, p. 315).

Farmers, in turn, insisted that everything of a demoralizing nature be abandoned: better to "have no fairs, and pay no premiums, if to do so we are compelled to resort to mean things" ("Managing County and State Fairs," 1874, p. 4). Men who found their moral objections dismissed were not long in doing the math needed to reveal the fallaciousness of arguments about the necessary evils that had to be endured for the greater good. Prize purses for horse races formed, by far, the greater portion of the fairs' premium competition expenses (Comstock, 1876). Was it not hypocritical for fair managers to allocate half of the prize money to good-for-nothing fast horses while cattle—a solid agricultural pursuit—were taxed with entry fees? Besides, what sort of agricultural good accrued from what happened inside the fairgrounds? In what way was horse speed useful for pulling crop cultivators and farm wagons? How, exactly, did the wonders of a snake show instill into the farm boy the desire to be an improving farmer? What little educational good the agricultural displays of the fair might have achieved was dwarfed by the horse races, the sideshows, and the crowds. Was it any wonder that the people in attendance took "the Fair more for a show or a place of amusement" than a source of valuable information related to the pursuit of their calling ("Dubois County Report," 1872, p. 310)?

Farmers were not so gullible as to think that improving agriculture figured largely into townsmen's interest in hosting fairs. Town boosters boasted of their ambitions to attract prospective migrants through fair-hosting, and, no one doubted that from the fair-going crowds the towns' merchants reaped the benefits. If these were the real aims, then farmers had some tips for improving the county fair's effectiveness. Was a growing population "more important to a country than any other thing?" Why not cut to the chase? Fair managers could offer a prize to the family that displayed the largest number of children of both sexes, with "size, healthy appearance, and good behavior" taken into account (T., 1874, p. 2). For drawing crowds to the county fairs, one farmer heard that Spanish bull fights were "great fun" with plenty of blood, gore, and excitement. Another suggested that agricultural societies might revive "the athletic games of the old Romans" as well. If they did, they could encourage the development of muscle and endurance "in the most noble of all animals—man" (Jake, 1879, p. 4).

The suggestions were meant to be ridiculous, the message serious. The "necessary evil" argument had been "a very successful rod for whipping in any and all who betrayed only conscientious scruples" but no longer (Templin, 1875, p. 2). Farmers had their fill of seeing town-boosting ambitions carried out in the name of agricultural progress. Exposure to demoralizing influences was not necessary, not for agricultural improvement, and not even for fair-hosting. The New York State Fair of 1874 had "wholly ignored horse racing, and the receipts were more than $10,000 more than usual." The Kansas State Fair enjoyed a similar experience. When the racers, gamblers, and hucksters were barred, the respectable elements of town and country turned out for the fair ("County and State Fairs," 1875, p. 4).

If Indiana's fair managers needed a few object lessons from places closer to home, they were not hard to find. While the county agricultural societies were going the way of the big fair and fast horse, farmers began creating their own independent agricultural clubs. Others joined with villagers to form township clubs. At least 34 township clubs were in operation by 1872, and the number of similar clubs proliferated with the rise of the Grange (State Board of Agriculture, 1872). Fair-hosting was not their primary mission, but displaying grains, vegetables, and fruits at harvest-time was a logical extension of their discussion meetings. Events of this character might not be considered fairs, but they had a remarkable tendency to grow into full-day events. Free of admission fees and morally objectionable features, the township fairs' significance might have remained strictly local and limited, had they not emerged (in part) as a protest against the degeneracy of the county fair.

The farmers of Brown Township formed an independent local club in 1867. Two years later, since the county agricultural society had "gone wild with the fast horse," they "concluded to try the experiment" of hosting a one-day, strictly agricultural fair. A "small show" at the schoolhouse was expected; by two o'clock in the afternoon 517 entries from the farm, garden, and kitchen were registered (Harvey, 1869, p. 77). With three years of experience, the township fair of the Honey Creek Agricultural and Horticultural Society grew into a three-day event. No "timing horses" were present, but the farm families came in droves, putting on display more than 250 animals and more than 100 products of the soil; the town-dwelling contingent contributed nearly 200 manufactured products ("Honey Creek," 1868, pp. 283–284). In Danville, the prize offer of a purebred lamb for the best half-bushel of potatoes, along with an open invitation to display all kinds of produce, was enough to generate a "large crowd" with a "show of farm products that beat [the Hendricks] County Fair" (W. H. S., 1869, p. 248). Cosponsored by farmer clubs and village residents, the township fairs provided conclusive evidence that town and country could come together for progress at an annual fair.

Held on schoolhouse lots, farms, and village commons, the township fairs were (on the whole) improvised events from year to year. Lacking a permanent institutional infrastructure, they depended upon the voluntary cooperation and enthusiasm of the participants. Just how many existed in any given year is impossible to gauge. As independent clubs, they were not obliged to report to the State Board of Agriculture, but a few dozen made regular reports anyway. Some made occasional contributions to the *Indiana Farmer* describing how they went about holding "a real agricultural fair" without the aid of "horse racing, shooting galleries, or monstrosities in human shape" (Danville, 1877, p. 5). Without exaggerating, their organizers could claim that "a well-managed township fair will do more good than many of our county fairs." Located closer to the farmers' homes, the township fairs were "more likely to reach the masses of the people." They were also far better "calculated to induce a spirit of friendly emulation in a community," since farmers who refused to attend demoralizing county fairs would attend township fairs (Clark, 1877, p. 3).

Improving farmers did not need township fairs, but they wanted them. At its best a fair could provide forms of agricultural education that farm papers only dimly represented. No line-drawing could inspire like a Shorthorn bull in the flesh; no amount of reading could match the learning derived from talking with other farmers. A good township fair offered something for everyone, from the successful farmer who hoped to sell the offspring of his improved livestock to the boys and girls who displayed their own animals, plants, and schoolwork. After weeks of long hours working the harvest, the entire farm family welcomed the fair as a holiday celebration. Neighbors and friends to see, news and gossip to catch up on, innovations in household goods and farm tools to inspect, all these and more were benefits of a township fair. It was everything that a county fair was supposed to be, a family friendly event for informal learning and socializing. By meeting the popular demand for fairs that were morally acceptable and practically useful, the township fairs deprived county fairs of what they needed most: agricultural legitimacy and farmers. With a low-budget harvest festival close to home, farmers had little reason to push for reform in the county fair.

Harmonizing Town and Country at the Fair

Avoiding what they disliked about civic life, rather than engaging in efforts to improve it, was farmers' natural inclination. With the rise of the Grange that tendency might have translated easily into a full-scale boycott of county fairs. The Grange was a temperance organization, after all, and some farmers condemned horse racing itself as immoral. The local grange also performed functions similar to those claimed by county fairs. Like the independent farmer clubs, granges held fall harvest displays. Many of the granges hosted demonstration trials of agricultural implements. Summer picnics were large social gatherings that featured agricultural displays and speakers on agricultural topics. Most of these events were open to families that did not belong to the Grange. For farmers who objected to the county fair's degeneracy, events of this sort readily took its place (Coffin, 1873; Wilson, 1878).

Social events sponsored by the Grange could be turned into large fairs with relative ease. Only three things were needed: the collaboration of several neighboring granges for resources and manpower, a few men willing to take charge, and someone with acres to spare for hosting the event. Cooperation from nearby townsmen was helpful but not essential. A two-day "jubilee" in Jefferson County was sponsored entirely by the grangers. The Jefferson County Grange Jubilee proved to be a resounding success. Within three years, it put the county fair out of business (Wood, 1877; "Jefferson County Report," 1883).

At its peak in the mid-1870s, the Grange had the farmers, resources, and organizational capacity to drive nearly all county fairs to ruin. To a considerable degree, the local fairs of the granges, independent clubs, and townships had this effect. By 1879, county fairs in only 51 of Indiana's 92 counties were left standing (State Board of Agriculture, 1879, pp. 351–352). This was not the outcome grangers desired. They did not want county fairs to fail and they did not try to create fairs that rivaled the county fairs in scope and scale. Generally speaking, large granger fairs were last-straw measures, created only after agricultural societies proved unwilling to reform. This was the situation that gave rise to the Jefferson County Grange Jubilee. Tellingly, after the county fair went under, the grangers moved their jubilee into the fairgrounds and it became the new county fair.

Grangers did not want separate fairs for town and country; nor did they want a host of small township fairs. Both types of alternative to the county fair could have been established easily on a permanent basis. Instead, grangers put their energies into reforming the county fairs. Their approach to the county fair was largely the same as it was to electoral politics and economic cooperation. The grangers tried to channel farmers' complaints toward moderate reform, while restraining farmers' inclinations toward extreme measures. Grangers had sufficient appreciation for what established institutions—whether political parties, railroads, free market competition, or county fairs—could be at their best to want to preserve them. They had just enough of the crusading spirit to think that they could compel town-dwellers to restore those institutions to true principles. And, as a combination of the most prominent farmers and those living closest to town, the Grange leadership had sufficient civic spirit to perceive how accommodating competing aims could further the interests of both town and country.

Civic spirit, reform zeal, and conservative inclinations go a long way toward explaining why grangers insisted that farmers try to restore the county fair to first principles. The remainder of the explanation resides with *Indiana Farmer* editor (and leading fair critic) J. G. Kingsbury's admission that the county fair was "a school of too much value to the farmer to be abandoned without weighty reasons" (Kingsbury, 1874, p. 4). Kingsbury did not need to explain to his readers why only the county fair could be such a powerful educator. Everyone knew that geographic scope gave the county fair a distinct educational advantage over local township and club fairs. In every county, only a few men had the wealth necessary to import livestock and maintain purebred herds. They could not bring their fine animals to dozens of township fairs. Therefore, it was essential that the farmers come to the county fair for their instruction and emulative example.

The prominent farmers who sought to provide a first-class livestock curriculum at the county fair were not disinterested benefactors. In the calls for fair reform that appeared in the agricultural press, farmers' self-interest was stressed as the main reason for taking an interest in the fairs' management. Properly understood, the appeal to self-interest meant that farmers could expect practical and educational benefits from fair-attending. With horses circling the track, it is somewhat difficult to imagine a farmer proceeding through the fair "with note book in hand, taking down all the new and valuable facts" ("County Fairs," 1871, p. 336). Nevertheless, there was some excellent agriculture on display from which the farmer could gather practical truths to apply to his own farm. Foremost among those truths, in age of grain and hog farmers, was the earning potential of improved livestock. Ambitious farmers were seeking to enlarge and improve their herds, but they were unwilling to attend demoralizing county fairs. Holding livestock auctions instead of horse races was a leading suggestion for fair reform (Maze, 1874, p. 1).

Both the would-be livestock sellers and purchasers were members of the county agricultural societies, but hitherto, only those seeking to sell had been active members. The Grange put farmers in the habit of attending meetings. Agricultural societies were faced with a simple choice: concede to farmers' demands or have reform thrust upon them by the force of numbers. The Delaware County Agricultural Society experienced what was, in effect, a hostile takeover. A name change, "an entirely new set of directors," and a complete ban on horse races and "clap-trap devices" followed ("Delaware County Report," 1875, pp. 249–250). Sullivan County, too, was compelled to dispense with the "agricultural trot." In its stead appeared a "liberal premium" for the grange making the "the best display of garden, field, and orchard products." The campaign to conduct the county fair "on a strictly moral basis" met determined opposition, but fair-goers left the grounds "satisfied that they got the worth of their money in what they saw" ("Sullivan County Report," 1877, pp. 221–222).

Elsewhere, public sentiment and depleted treasuries persuaded agricultural societies that a "return to first principles" was prudent. With the farmers staying home, losing $10 in gate receipts for every $1 gained from the hucksters and "gambling fraternity" was a powerful argument. Fair managers had little choice but to appeal "to the community to give them a chance to redeem their blemished reputation." Being considered the county's "favorite institution" brought its own gratification, but closing the fair with at least a "small balance in the treasury" was essential ("Fayette County Report," 1874, pp. 170–171). Experience proved that without "good feeling existing between the management of the fair and the grange element," a county fair would have great difficulty making ends meet ("South-Eastern District Report," 1879, p. 288).

Obtaining farmers' participation, however, created a new set of problems. Agricultural societies were confronted with the challenge of guarding public morality. Thousands of people thronged the fairgrounds, hundreds of exhibitors and competitions required attention, and there was seldom a shortage of sideshows clamoring for admission. Given the small number of people charged with operating the county fair, maintaining order furnished enough trouble. It was compounded by the fact that some of the people in charge were not committed to the cause of moral reform. The farmers made their voices heard at the agricultural society meeting, but the county seat's merchants and businessmen ran the fair.

With leading townsmen deciding which sideshows could be permitted onto the grounds, there was bound to be some slippage in enforcing prohibitory regulations. Yet, it is hard to explain how it came to pass that a "large number of gamblers" plied "their nefarious and unlawful trade without molestation" on the Shelby County fairgrounds ("The Shelby County Fair," 1879, p. 1). In Howard County, the "betting was brisk" on the favorites, while a "score or more of gambling schemes" kept the young men entertained in between heats. The agricultural society "received $200 for the privilege of swindling on the grounds" ("The Fairs: Kokomo," 1879, p. 1). The "gentlemen of character and standing" in Dearborn County displayed the same inclinations. A committee of five appointed to police the grounds decided that a "pool wheel" (wheel of fortune or roulette) "was no 'species of gambling' ... [and] ... that whisky, colored to resemble red lemonade, was 'no species of spiritous [sic] or malt liquors' ... Thus did an agricultural fair, a promised occasion of sobriety and chastity, run to the resemblance of a drunken orgie [sic]" (Kerr, 1876, p. 4).

The Vigo County Agricultural Society almost managed to conduct a strictly "moral affair," but for some inexplicable reason, "the Board granted the privilege of erecting a stand upon the ground, of sufficient size in which to carry on dancing." When the nature of the activity inside the tent was discovered, fair managers refused to prohibit the display of women's breasts and legs. Instead, "no dancing was allowed after night" ("Vigo County Report," 1875, p. 308). On his travels throughout the State, J. G. Kingsbury's observations raised "strong suspicions" that Vigo County's dancing stand was no aberration. He had "indulged the hope that a reform had taken place among the managers of the fairs," but found demoralizing influences to be widespread ("Gambling at Fairs," 1878, p. 4).

Lax enforcement heightened farmers' determination to see that the next fair was completely reformed. The horse races had to go. "Where horse racing is made a special feature at fairs, it almost always happens that chance games, chuck a luck, cards, roulette, lottery wheels and the like, are freely licensed" ("Gambling at Fairs," 1878, p. 4). Horse races brought large crowds; crowds brought traveling sideshows and gamesters. The temptation to profit from debauchery was too much for agricultural society members to resist; experience convinced farmers of that. The result was an ongoing struggle between the grange element and the town element, between those who opposed horse racing "from purely conscientious motives" and those who "advocated it from a love of the sport, or because they considered it a [financial] necessity" ("Huntington County Report," 1876, p. 135).

The struggle for control over the agricultural societies took place on two fronts: at the annual meeting for electing officers and at the county fair. Between them enough reversals of policy were generated to create a cycle of self-fulfilling prophecies. Disgusted with a fair run as "a speculating scheme," the alliance of respectable farmers (and the town's moral crowd) would sweep "a new set of officers" into command and resolve to "make it *Our* fair" ("Whitely County Report," 1883, p. 334). Deprived of horse races, the next fair was almost certain to have a smaller crowd, thereby lending support to the claim that hosting races was a financial necessary. At the next annual meeting, the "lawyers, county officers, merchants, clerks, grog-shop loafers and others, not agriculturists turned out *en masse,* and voted into management officers of their liking" ("Controlling County Fairs," 1877, p. 4). The horse races and gamblers would return to the next county fair to be encountered by respectable farmers who, based on the previous year's experience, thought they were bringing their families into a strictly agricultural fair. The cycle from agricultural fair to degeneracy and back might swing from season to season, or it might take a few years before farmers (convinced that reform had been achieved) failed to show up at the appointed time for electing officers. In the meantime, who could know what to expect from the county fair?

Something had to give. The year 1877 marks a decisive turning point. That year, the Indiana General Assembly granted agricultural societies permission to borrow up to $10,000 from the county treasury to pay for improvements to the fairgrounds (already made or desired). A majority of the county's voting residents had to approve ("Act authorizing allowances," 1877). Without the aid of leading farmers, the legislation could not have passed; in the counties there was little chance of getting referenda approved unless farmers were persuaded that genuine reform efforts would follow the customary proclamations of good intentions. If prominent farmers were willing to give agricultural societies the chance to win public aid, there were townsmen ready to accept the inevitable conditions that would accompany it. Leading farmers and town merchants had come to the conclusion that their interests in successful county fairs were mutual.

The deal (where struck) was straight forward. The demands of financial necessity would be honored by retaining the horse race, but prohibitions against degrading sideshows and gambling were to be strictly enforced. With respect to horse racing, moralizers who had conscientious scruples against it would have to be content with attending fairs that had less of it. "Talk about pumpkins and potatoes, as much as we love them, and would gladly give them the first place if possible, but [we] are compelled to yield. They will not draw like the horse" ("Porter County Report," 1884, p. 290). Fair managers had long made that point, but leading farmers (faced with diminishing premium awards) and aspiring farmers (desirous of access to the best livestock) were now ready "to try racing once more as an experiment" ("Huntington County Report," 1878, p. 232). The experiment's details varied. Certain days might be appointed as race day; prize purses might be reduced to discourage professional horsemen (and gamblers) from attending; or, the number of races reduced to one or two per day. Whatever the method, the object was the same, to diminish horse racing's status as the feature attraction and to elevate to prominence the "many things connected with agriculture that present stronger claims to usefulness" ("Randolph County Report," 1877, p. 216).

Improvement-minded farmers were "half inclined to admit" that horse racing was, in fact, a "necessary evil" if the county fairs were to be sustained ("Horse Racing Fairs," 1877, p. 4). Farmers drew a hard line around gambling, drinking, dancing, and swindling sideshows though. Agricultural societies were ready to give up license fees in exchange for farmers' participation. Purging the county fair of immorality achieved the desired effect. "The farmers came in from all directions, and appeared to be glad that they were there" ("Grant County Report," 1879, p. 306). Men who could not attend could rest easy, knowing that they could "send their wives and little ones, assured that they will be unmolested and safe from annoyance." A fair visitor from a few years back would have rubbed his eyes in astonishment, but the county fair had become an "orderly, quiet, peaceable place of resort" ("Gibson County Report," 1880, p. 273). The claim was overstated, but it was undeniably the case that a marked change had come over the county fair. With a slight adjustment to account for the intergenerational change in public sentiment regarding horse racing, the county fair was almost restored to the first principles that had been established for it in the 1850s.

One should not conclude that farmers' efforts to reform the county fairs were an overwhelming success everywhere. In some places, the struggle among competing elements of the community continued through the 1880s. After several years of completely reformed fairs, the Delaware County Agricultural Society reintroduced racing on a limited scale; a few years later, with wagers "on every hand from the horse race down to the game of nine-pins," its 1884 fair was pronounced a "gambler's paradise" by the "better class of citizens" (E. J. T., 1884, p. 11). That same year, Ripley County witnessed a fair that was a "libel on agriculture"; naturally, farmers campaigned for a change in the management (Spencer, 1884, p. 1). Saddled with debt, the Tippecanoe County Agricultural Society had the temerity to appeal to "the temperance men to lay aside their principles and whoop and hurrah for the 'get alongs' and the beer stand" for the greater good achieved by the county fair. There was less chance of that than of winning a spin at the pool wheel: "far better to suffer the loss of the fair, notwithstanding its great renown and many benefits to town and country" (G. W. T., 1884, p. 8).

In places where moral principles and material interests could not be reconciled county fairs failed. In the 1880s, roughly one-third of Indiana's county fairs were abandoned as a result of some combination of the moral indifference of leading townsmen and the civic indifference of farmers (State Board of Agriculture, 1889). Their fate marked the limits of shared commitments across town and country in Indiana.

Putting True Progress into the County Fair

In the two-thirds of Indiana's counties where moral reform was sustained, the county fairs took on the character of (for want of a precise term) annual events that brought town and country together to celebrate their mutual progress. Two innovations—pioneer days and school exhibits—symbolized the change. Both were launched into prominence by their presence at the State Fair, the fair that took the most abuse for its degeneracy. At the urging of John B. Dillon, one of Indiana's earliest historians, the State Board of Agriculture agreed to host the first convention of the Pioneer Association of Indiana in 1878. All pioneers (at least 70 years old and with 40 years of residence in Indiana) who "laid the foundation for this great and growing state" were invited to attend the State Fair, free of charge, for a formal ceremony in their honor ("State Pioneer Convention," 1878, p. 376). The idea was received with enthusiasm around the state. County fairs set aside a day to honor the "old settlers," and to the formal ceremony were added features well suited to a fair: log cabins and "old relic" departments. The displays featured the senior members of the community regaling "the young and rising generation" with "old time stories, displaying old relics, baking Johnny cakes, breaking and sketching flax after the old style," and performing other tasks associated with the settlement of Indiana's frontier. One might wonder what the countryside's farm families who were living yet in the old style thought of all this, but the exhibits' popularity made it clear that the spirit of improvement had come of age in Indiana ("Wayne, Henry, and Randolph District Report," 1888, p. 295).

Displaying old relics and old-timers was one way to measure how much progress had been achieved in Indiana in the span of a single lifetime. Putting schoolwork on display revealed the potential and determination of the rising generation to carry true progress forward. With some difficulty, the State Superintendent of Schools attempted to get a competition among city schools initiated at the State Fair. Only the Terre Haute schools responded to the initial call in 1874 (State Board of Agriculture, 1874). Continued solicitation brought exhibits from the state institutions for the deaf, dumb, and blind, Purdue University, and a number of high schools and common schools from larger cities and towns. To make the competition fair, and to encourage friendly rivalry among similar communities, separate categories were created for colleges and high schools, city schools (population exceeding 5,000), graded schools in towns (population less than 5,000), and district schools. Arranged according to the home communities represented, the clusters of outline maps, line drawings, penmanship samples, copy books, examination papers, and botanical and geological specimens had little to do with agriculture. They had everything to do with progress. Eager to ensure that their cities and counties were well represented at the State Fair agricultural societies began sponsoring their own county competitions ("Educational Department of State Fair," 1878).

To some extent, the schoolwork exhibits were natural outgrowths of the Fine Arts and Natural History displays that were present in most fairs. Encouraged by special premiums from leading townsmen, these exhibits typically included oil and watercolor paintings, line drawings, and penmanship samples, as well as preserved plant specimens and fossils. Somewhat out of place in an agricultural fair, artifacts of this sort were found in the Floral Hall or the Miscellaneous Hall. Surrounded by flowers, one can readily appreciate how the exhibits inspired admiration for beauty, good taste, and scientific knowledge. Among the oddities of all kinds—as well as an endless array of mechanical contraptions and consumer goods—found in the Miscellaneous Hall one cannot help but wonder at their effect upon fair-goers. The turn to schoolwork displays put some order into the chaos and bolstered efforts to separate the fair's serious educational features from those that possessed entertainment (and commercial) value only. At the same time, the schoolwork broadened the fair's usefulness by making children's work part of the premium competition.

Characteristically, when the school exhibit was introduced, the city graded schools and normal schools were the principal contributors, but the competitions were open to the district schools of the countryside, too. Before long, the "persevering effort" of the county school superintendents was rewarded. The promise of "free tickets" awakened a "lively enthusiasm" between teachers and pupils. Some 5,000 schoolchildren in Steuben County "were made happy in a free show" and, in exchange, put on a display of schoolwork that was "highly creditable" ("Steuben County Report," 1887, p. 232). In Hancock County, "almost every district in the county was represented" when the children "formed in line at the Court House and marched thirteen hundred strong to the fair grounds, with mottos, flags and banners flying" ("Hancock County Report," 1887, p. 196). Horse races were conspicuous by their absence on children's day, but attendance and gate receipts proved favorable, "notwithstanding the fact that some of the oldest looking 16-year-old chaps to be found in the coon belt passed in free" ("Howard County Report," 1886, p. 247).

The schoolwork exhibit did something the fast horse could never accomplish: it brought the town's moral crowd and the countryside's respectable farm families into the fairgrounds. The secretary of the Wells County Agricultural Society did not mince words about it: "We have created a class or 'Division of Education' in our county fair, for the purpose of stimulating an interest in the cause of education and awakening a new interest in the fair" ("Wells County Report," 1877, p. 233). The strategy worked, for the schoolwork and old relics appealed to a broad swath of the community across town and country. At the same time, they put the ingredients of true progress at the heart of the county fair. In these respects, as a social gathering and site for learning and (non-offensive) amusement, the reform of the county fair was complete in the mid-1880s.

There was, however, one thing missing that prevented the fair from encompassing the "best work of the county." Its "business side" was still largely deficient of local agriculture ("Lake County Report," 1881, p. 279). Even county fairs located in premier agricultural regions witnessed a greater number of exhibits in the categories of natural history, fine arts, and mechanical than in agriculture. Growth in those categories was a heartening sign of progress, but it revealed that the county fair's competitions had not been revised sufficiently. Local farmers were attending the reformed county fairs, but they were not joining in the competitions.

The problem was the same as it had been since the late 1860s. Agricultural societies offered large premium awards to attract the best livestock to their fairs. After the county fairs were declared open to the world, professional exhibitors (who traveled from place to place displaying the same animals) dominated the competitions. The livestock were impressive but the competition was a "tame affair," with hardly anything resembling "adequate competition" ("Clinton County Report," 1886, p. 232). Local farmers showed little interest in participating, having "learned from past experience that they [could] not successfully compete" against the "traveling caravans" ("Pike County Report," 1888, p. 269). Hailed as the embodiment of progress by fair-boosters of the 1870s, the "man with the show herd" came under mounting criticism in the 1880s ("The Man with the Show Herd," 1887, p. 5). A growing cohort of farmers wanted to take part in the county fair's competitions. At great expense and effort, they had improved their cattle, horses, and hogs. They now insisted that the "public spirited men" who were making "the stock business profitable" for Indiana's farmers receive their share of the honor as "public benefactors" ("Elkhart County Report," 1881, p. 259).

The men grumbling against the traveling caravans of showy livestock had good reason to be sore. They—not the professional exhibitors—were responsible for improving local herds. As their counties' leading farmers, they were importing purebred livestock and selling offspring to local farmers. Their purebred bulls, rams, and boars were performing stud service to upgrade the quality of other men's farm animals. Through face-to-face conversations, they were educating common farmers about the merits of particular livestock breeds for local conditions and specific marketing purposes. Professional exhibitors did none of these things. The question, therefore, was overdue: if agricultural fairs "are the schools from which proper education ought to be disseminated," how should the competitions be designed to "exert the greatest amount of good" (Louder, 1878, p. 1).

Opinion in the 1880s fell squarely on the side of stimulating home-grown competition. In county after county, improving farmers insisted that the competitions be restricted to local residents. The litmus test for a "good fair" ceased to be a small show of the finest agriculture and became a full show furnished by "plenty of small exhibitors" (Stahl, 1884, p. 1) Having most exhibits furnished by "amateurs," or declaring the fair to be "strictly a county organization," became badges of honor. Agricultural societies that adopted this course "struck the popular chord" and found friends among farmers who had been the staunchest opponents of the county fair ("Carroll County Report," 1887, pp. 175–176).

One final reform in the agricultural department was necessary. Livestock were essential to a fair, and they were increasingly a part of farmers' moneymaking business. No matter how fine or how profitable though, livestock were not agriculture proper: grain, vegetables, and fruits were "the very foundation of correct farming" ("Elkhart County Report," 1883, p. 283). These crops—the work of small-scale farmers—had never received much honor at county fairs. Agricultural societies were compelled to admit that their awards showed "too much discrimination" in favor of cattle and horses "by organizations professedly agricultural." It was time to give more prize money to the "tillers of the soil" ("Vigo County Report," 1887, p. 241). What better way to "stir up a laudable emulation amongst farmers," to encourage "the use of fertilizers and improved tillage," and to challenge every man "to produce as much on 10 acres as he now does on 20" ("Pike County Report," 1885, p. 320)? The idea was about as original as *Poor Richard's* "little farm well-tilled," and a premium award is a small token. It revealed something important though. The townsmen in charge wanted more local farmers to participate in the county fair's premium contests. For the first time in their history, with a broadened pool of farmers involved in the competitions, agricultural societies could claim truthfully that their fairs exhibited "the steady progress our people are making in all the industries" of the county ("Knox County Report," 1887, p. 206).

As for agricultural societies that could not make this claim, there were too many county fairs that thrived on useful improvement for fair directors to claim that they could not "get the people out" unless fast horses and gambling were feature attractions. To admit that, J. G. Kingsbury thought, was to pay a "very poor compliment to the intelligence and morality" of the community: a degenerate fair was "a stigma on the character and good sense of the agriculturalists" (Kingsbury, 1884, p. 8). In the mid-1870s, when the typical county fair was a town-boosting organ, the *Indiana Farmer* editor would not have made such a statement. A decade later, the county fair belonged to the countryside as well as the town. If it was not a proper agricultural fair, the county's farmers lacked the commitment to self-improvement and the home county pride that were necessary to make it one. Both qualities of character were expected from men who claimed to be worthy farmers and citizens. The annual fair, to be sure, was not a direct reflection of a people's character, but it did yield a sightline onto J. G. Kingsbury's broader point. The steady growth of manufacturing notwithstanding, Indiana remained an agricultural state. For true progress to occur, the right kinds of farmers were needed, and they had to make their influence felt in their civic communities.

Reflections on Civic Learning

Improving agriculture formed the public good rationale for hosting fairs. In the 1870s, though, the primary purpose was to make money for the county seat's leading businessmen, whether directly from the crowd's spending during fair-week or indirectly from potential investment opportunities and real estate sales. Over the long term, a secondary purpose might be achieved, the diversification and growth of local industries. Claiming that horse races (and all the rest) were necessary evils to be endured for the greater agricultural good was hypocrisy. Townsmen and farmers knew it.

The county fair of the 1870s did not reflect the ethos of true progress. The economic growth ambitions were too excessive and public morality was sacrificed for material gain. The economic growth orientation itself was misdirected: it looked abroad for the sources of improvement (investment capital and potential migrants) rather than locally to the enterprise of the county's residents. Efforts to encourage the people's self-improvement by diffusing useful knowledge were feeble, at best. To the extent that the county fair reflected the community's interests, its area of concern was confined to the county seat. Local agriculture had almost no place in the competitions, and the livestock were intended for show, not the instruction of the countryside's farmers. The rural community's premier civic event yielded ill-fitting answers to the civic questions it raised every year: What kind of community is this? What do we value? Indiana's improvement-minded farmers, at least, thought so.

The county fair was supposed to be a testimony of the community's character. As a public event, its purposes and program were things upon which the community should have agreed. That was not the case in the mid-1870s, for there was not a cohesive sense of the community on economic development and the fair's role in it. Leading townsmen were pursuing economic development by any means necessary. Confronted with boosterism and debauchery, farmers boycotted the county fair. Prevented by the Grange leadership from forming an alternative system of township fairs, farmers took their complaints (and their votes) into the annual elections of the county agricultural society. The tug of war over the agricultural society's management pitted the countryside's leading farmers against the county seat's leading businessmen. The events of the county fair itself became key ingredients in disputing what aims the community ought to pursue as a shared agenda, as well as what was justifiable in the name of the public good.

Support of the degenerate fair was found solely in the county seat; opposition to it was dispersed throughout the county. It made sense, therefore, for people to perceive fair reform as the countryside's crusade to purify the town's morals. The solution, however, did not reside in mobilizing farmers to close down the county fairs. One might say that cooler heads prevailed. Or, one might say, instead, that the controversies provided opportunities for a portion of farmers and town-dwellers—those who thought of themselves as the respectable portion of the community—to learn something about each other. They shared common cause against the county's elite who profited from popular vice and the degraded masses who rightly took the fair for a gambling festival.

Recognition of common cause across town and country was not quite the level of shared commitment that might qualify as the sense of the community. In roughly one-third of Indiana's counties, a solid commitment to fair reform could not be built and sustained by respectable townsmen and farmers. In every county, the poorest farmers ignored the fair entirely or joined the towns' working poor in the revelry. Nevertheless, by the 1880s, a coalition in favor of the useful and the good did take shape across much of Indiana. Immorality was purged, schoolwork and the frontier heritage were given prominent places, and practical agriculture was elevated to where it properly belonged in a celebration of rural community.

Respectable farmers and townsmen made their priorities the heart of the county fair. They held the balance of power in the rural community. They were determined to see that public institutions like fairs represented their interests and principles, not those of the county's elite or those of the degraded masses. In its own way, the reformed county fair reflected traditional republican ideals, a version based on individual merit, morality, and the leadership of those who had a solid stake in the community (if not large quantities of landed property). If nothing else, where town and country came together in the 1880s, the county fair signified true progress. For the men who controlled it, that was the equivalent of the common good.

References

Act authorizing allowances in aid of agricultural associations. (1877). *Indiana Laws* (50th Session), 3.

Carroll County report. (1887). In *Annual Report of the Indiana State Board of Agriculture*. Indianapolis, IN: State Board of Agriculture.

Cass County report. (1874). In *Annual Report of the Indiana State Board of Agriculture*. Indianapolis, IN: State Board of Agriculture.

Clark, A. (1877). Township, or grange fairs. *Indiana Farmer, 12*(38), 3.

Clinton County report. (1886). In *Annual Report of the Indiana State Board of Agriculture*. Indianapolis, IN: State Board of Agriculture.

Coffin, E. (1873). Letter from Hancock County. *Indiana Farmer, 8*(7), 3.

Comstock, H. (1876). Racing at our fairs. *Indiana Farmer, 11*(12), 2.

Controlling county fairs. (1877). *Indiana Farmer, 12*(4), 4.

County and state fairs. (1875). *Indiana Farmer, 10*(9), 4.

County fairs. (1871). *North Western Farmer, 6*(9), 336.

Danville. (1877). Wheat-sowing—a model fair. *Indiana Farmer, 12*(41), 5.

Deal, S. (1873). A suggestion to farmers. *Indiana Farmer, 8*(11), 33.

Decatur County report. (1873). In *Annual Report of the Indiana State Board of Agriculture*. Indianapolis, IN: State Board of Agriculture.

Delaware County report. (1875). In *Annual Report of the Indiana State Board of Agriculture*. Indianapolis, IN: State Board of Agriculture.

Don't go in debt. (1874). *Indiana Farmer, 9*(5), 4.

Dubois County report. (1872). In *Annual Report of the Indiana State Board of Agriculture*. Indianapolis, IN: State Board of Agriculture.

The duty of fair managers. (1870). *North Western Farmer, 5*(1), 7.

Educational department of state fair. (1878). *Indiana Farmer, 13*(15), 2.

Elkhart County report. (1881). In *Annual Report of the Indiana State Board of Agriculture*. Indianapolis, IN: State Board of Agriculture.

Elkhart County report. (1883). In *Annual Report of the Indiana State Board of Agriculture*. Indianapolis, IN: State Board of Agriculture.

E. J. T. (1884). Notes from Delaware County. *Indiana Farmer, 19*(35), 11.

The fairs: Kokomo. (1879). *Indiana Farmer, 14*(37), 1.

Fayette County report. (1874). In *Annual Report of the Indiana State Board of Agriculture*. Indianapolis, IN: State Board of Agriculture.

Gambling at fairs. (1878). *Indiana Farmer, 13*(28), 4.

Gibson County report. (1871). In *Annual Report of the Indiana State Board of Agriculture*. Indianapolis, IN: State Board of Agriculture.

Gibson County report. (1880). In *Annual Report of the Indiana State Board of Agriculture*. Indianapolis, IN: State Board of Agriculture.

Grant County report. (1879). In *Annual Report of the Indiana State Board of Agriculture*. Indianapolis, IN: State Board of Agriculture.

G. W. T. (1884). Tippecanoe County agricultural society. *Indiana Farmer, 19*(2), 8.

Hancock County report. (1887). In *Annual Report of the Indiana State Board of Agriculture*. Indianapolis, IN: State Board of Agriculture.

Harvey, J. L. (1869). The Brown Township agricultural and horticultural society. *North Western Farmer, 4*(4), 77.

Hazlet, J. (1874). How to make the farm pay. *Indiana Farmer, 9*(8), 4.

The Hendricks County (Ind.) fair. (1868). *North Western Farmer, 3*(10), 222–223.

Hendricks County report. (1869). In *Annual Report of the Indiana State Board of Agriculture*. Indianapolis, IN: State Board of Agriculture.

Honey Creek agricultural and horticultural Society. (1868). In *Annual Report of the Indiana State Board of Agriculture*. Indianapolis, IN: State Board of Agriculture.

Horse racing fairs. (1877). *Indiana Farmer, 12*(38), 4.

Howard County report. (1886). In *Annual Report of the Indiana State Board of Agriculture*. Indianapolis, IN: State Board of Agriculture.

Huntington County report. (1873). In *Annual Report of the Indiana State Board of Agriculture*. Indianapolis, IN: State Board of Agriculture.

Huntington County report. (1876). In *Annual Report of the Indiana State Board of Agriculture*. Indianapolis, IN: State Board of Agriculture.

Huntington County report. (1878). In *Annual Report of the Indiana State Board of Agriculture*. Indianapolis, IN: State Board of Agriculture.

Jackson, C. P. (1872). Conducting fairs. *North Western Farmer, 7*(8), 164.

Jake. (1879). A bull fight suggested for the state fair. *Indiana Farmer, 14*(9), 4.

Jasper County report. (1879). In *Annual Report of the Indiana State Board of Agriculture*. Indianapolis, IN: State Board of Agriculture.

Jefferson County report. (1883). In *Annual Report of the Indiana State Board of Agriculture*. Indianapolis, IN: State Board of Agriculture.

Jennings County report. (1874). In *Annual Report of the Indiana State Board of Agriculture*. Indianapolis, IN: State Board of Agriculture.

Kerr, M. B. (1876). Agricultural fairs—should they be abandoned? *Indiana Farmer, 11*(38), 4.

Kingsbury, J. G. (1874). State and county fairs. *Indiana Farmer, 9*(30), 4.

Kingsbury, J. G. (1884). Attractions at fairs. *Indiana Farmer, 19*(38), 8.

Knox County report. (1887). In *Annual Report of the Indiana State Board of Agriculture*. Indianapolis, IN: State Board of Agriculture.

Lake County report. (1881). In *Annual Report of the Indiana State Board of Agriculture*. Indianapolis, IN: State Board of Agriculture.

LaPorte County report. (1873). In *Annual Report of the Indiana State Board of Agriculture*. Indianapolis, IN: State Board of Agriculture.

Louder, C. (1878). Discussion. *Indiana Farmer, 13*(2), 1.

The man with the show herd. (1887). *Indiana Farmer, 22*(24), 5.

Managing county and state fairs. (1874). *Indiana Farmer, 9*(41), 4.

Maze, W. A. (1874). Suggestions on fairs. *Indiana Farmer, 9*(1), 1.

Neidhart, G. W. (1875). About rural economy. *Indiana Farmer, 10*(8), 6.

Old School Farmer. (1873). Management. *Indiana Farmer, 8*(12), 11.

Pike County report. (1885). In *Annual Report of the Indiana State Board of Agriculture*. Indianapolis, IN: State Board of Agriculture.

Pike County report. (1888). In *Annual Report of the Indiana State Board of Agriculture*. Indianapolis, IN: State Board of Agriculture.

Porter County report. (1874). In *Annual Report of the Indiana State Board of Agriculture*. Indianapolis, IN: State Board of Agriculture.

Porter County report. (1884). In *Annual Report of the Indiana State Board of Agriculture*. Indianapolis, IN: State Board of Agriculture.

Putnam County report. (1873). In *Annual Report of the Indiana State Board of Agriculture*. Indianapolis, IN: State Board of Agriculture.

R. (1876). Agricultural fairs. *Indiana Farmer, 11*(4), 3.

Randolph County report. (1877). In *Annual Report of the Indiana State Board of Agriculture*. Indianapolis, IN: State Board of Agriculture.

The Shelby County fair. (1879). *Indiana Farmer, 14*(37), 1.

South-eastern district society report. (1879). In *Annual Report of the Indiana State Board of Agriculture*. Indianapolis, IN: State Board of Agriculture.

Spencer, B. F. (1884). County fairs. *Indiana Farmer, 19*(34), 1.

Stahl, J. M. (1884). Prepare for the fair. *Indiana Farmer, 19*(32), 1.

State Board of Agriculture. (1872). List of agricultural societies. In *Annual Report of the Indiana State Board of Agriculture* (pp. 356–362). Indianapolis, IN: Author.

State Board of Agriculture. (1874). Public schools and the exposition. In *Annual Report of the Indiana State Board of Agriculture* (pp. 148–152). Indianapolis, IN: Author.

State Board of Agriculture. (1879). In *Annual Report of the Indiana State Board of Agriculture* (pp. 351–352). Indianapolis, IN: Author.

State Board of Agriculture. (1889). List of agricultural societies. In *Annual Report of the Indiana State Board of Agriculture* (pp. 400–404). Indianapolis, IN: Author.

State pioneer convention. (1878). In *Annual Report of the Indiana State Board of Agriculture* (pp. 375–378). Indianapolis, IN: State Board of Agriculture.

Steuben County report. (1887). In *Annual Report of the Indiana State Board of Agriculture*. Indianapolis, IN: State Board of Agriculture.

Stevenson, A. C. (1868). Our fairs. *North Western Farmer, 3*(12), 276.

Sullivan County report. (1877). In *Annual Report of the Indiana State Board of Agriculture*. Indianapolis, IN: State Board of Agriculture.

T. (1874). Premiums on children. *Indiana Farmer, 9*(10), 2.

Templin, L. J. (1875). Our coming fairs. *Indiana Farmer, 10*(9), 2.

Vigo County report. (1869). In *Annual Report of the Indiana State Board of Agriculture*. Indianapolis, IN: State Board of Agriculture.

Vigo County report. (1875). In *Annual Report of the Indiana State Board of Agriculture*. Indianapolis, IN: State Board of Agriculture.

Vigo County report. (1887). In *Annual Report of the Indiana State Board of Agriculture*. Indianapolis, IN: State Board of Agriculture.

Voice from Hendricks. (1868). Letter from Hendricks County. *North Western Farmer, 3*(12), 277.

Wayne, Henry, and Randolph district report. (1888). In *Annual Report of the Indiana State Board of Agriculture*. Indianapolis, IN: State Board of Agriculture.

Wells County report. (1877). In *Annual Report of the Indiana State Board of Agriculture*. Indianapolis, IN: State Board of Agriculture.

Whitely County report. (1883). In *Annual Report of the Indiana State Board of Agriculture*. Indianapolis, IN: State Board of Agriculture.

W. H. S. (1869). The great potato show. *North Western Farmer, 4*(11), 248.

W. H. S. (1874). A word to the order. *Indiana Farmer, 9*(5) 4.

Wilson, J. D. (1878). Dudley grange fair. *Indiana Farmer, 13*(41), 7.

Wood, D. (1877). Jefferson County Pomona Grange jubilee. *Indiana Farmer, 12*(38), 7.

8

Bringing Farmers into Town for a Strictly Agricultural Education

Agricultural Improvement's Civic Message and Indiana's Farmers

The Centennial Year of the Declaration of Independence furnished ample occasions for drawing civics lessons from history. Early in that year, two essays were printed in the *Indiana Farmer* that looked across time and geographic space (for the two dimensions were inseparable) to give Midwestern farmers the proper perspective on their place in American progress and their duties as farmer-citizens.

Taking stock of the Republic's first century, the first essayist told a familiar story about the "conquests of agriculture" for the American empire. Contending for "every inch of the ground with the forest giants," the pioneering farmer cleared the way for an "empire of fruitful nature east of the Mississippi," then "crossed over to the rich plains beyond" to subdue new regions. The well-known version of this story of Man's triumph over Nature gives pride of place to the pioneering farmer. Accounts penned by advocates of agricultural improvement in the nineteenth century did not. The reason? Men spread across the continent because they did not know how

Civic Learning through Agricultural Improvement, pages 177–201
Copyright © 2011 by Information Age Publishing
177

to farm in ways suited to civilization. Once "outward influences like war, commerce and stable government" made themselves felt, the pioneering farmer became an impediment. Handy with the axe and jumper plow, but ignorant of valid agricultural methods, the pioneering farmer got diminishing yields from deteriorating soils until he sold out and moved on, to repeat again the farm-making and soil-wasting process.

In contrast to the pioneering farmers of previous generations, the current generation of solid farmers had been released from the "thralldom of the past" by agricultural science. Chemists' discovery that "certain definite elements, in definite quantities, invariably go to make up certain plants" made it possible to farm with an eye to productivity and sustainability. From that fertilizing awareness, an "intellectual revival" began. Influenced by agricultural societies and newspapers, farmers broadened their range of commercial crops. In doing so, they stopped relying on Indian corn for cash income and diminished the "great moral problem" (actually two, soil depletion and whiskey drinking) caused by its cultivation. Last in order, but greatest in significance, came the raising of improved livestock. That change alone gave farmers more "independence" than "all other sources," for "instead of being dependent upon the uncertainties of a single calling" (grain) the farmer had "the chances of two." Now "firmly established in the business and life of the American farmer," the pursuit of practical knowledge was propelling society forward, "towards something far beyond anything we now know" (J. H., 1876, p. 6).

The second Centennial Year essayist offered a somewhat different retrospective. In the early years of the Republic, everything related to farming "tended in the direction of impoverishment." The farmer toiled hard "to procure a subsistence from month to month and year to year, but not to get rich." When his fields "failed to do justice" to his "wretched culture," more trees were felled until "there was no more land to clear." The farmer and the farm "grew poorer instead of richer." That kind of "small farming to get a living"—more than the rocks, hills, and climate—"discredited New England farming" for subsequent generations. Bequeathed a "heritage of debts" and "lands too poor to even promise them a living," farmers' sons abandoned their home communities and headed west.

Relocation to the Midwest "inaugurated a revolution." Its "illimitable prairies and level forest lands" were "rich beyond calculation." Accessible by rail and sea, the markets of the eastern seaboard and Old World beckoned for all the grain that farmers could produce. As a result of the right match of environment and economic opportunity, among the newer generation of farmers, the "fundamental idea of farming was changed." Farming was no longer viewed as "a mere makeshift," it was "a business to be undertaken for profit, a road to fortune and independence." Land-skinning cultivation (although "not yet altogether abandoned") was giving way, for the new farmer's "leading idea" was "to increase the value of his land, to make it better instead of worse." No other domain of American life was making such "great headway as in agriculture whether its business be considered or the character of the farming class." Committed to improving and profiting from their farms, farmers stood at the forefront of progress (H., 1876, p. 6).

Rehearsed countless times over the years, the commemorative essays' underlying message was clear and familiar. It had three basic parts. The first part posed a stark choice: educate or relocate. Soil fertility could not be sustained indefinitely under wasting cultivation. Choosing the West's broad acres over agricultural improvement's little farm well-tilled ideal only put off the day of reckoning. The second part promised prosperity. Farmers who chose to educate and remain on their home farms did not simply scratch a subsistence living from the soil. Instead (and unlike farmers of the past) they thrived from having access to a nearby home market and distant urban markets via the railroad. Meeting the demands of multiple markets through diversified agriculture guaranteed a stable income via two or three main agricultural specialties (and, as need or opportunity dictated, raising smaller portions of other produce). The last part of the message tied farmers' self-interest to the public interest. Agricultural improvement was upheld as the right thing to do for the good of the community. Men who followed its prescriptions were held in the highest esteem for making possible the progress of civilization or, less grandly, local economic development.

These were the lessons of experience, the historical-social kind that represents an intergenerational shift in normative expectations and the teaching of personal experience, that farmers were expected to learn to fulfill their duty to self and society. It was no accident that the *Indiana Farmer* placed commemorative essays about pioneer farming and the New England experience side-by-side. In its location, developmental stage, and in the quality of its farmers, Indiana stood between the two reference points. Hoosier readers could readily see themselves and their neighbors in the portrayals of praiseworthy farmers and soil-wasters.

Farmers who took the lessons of the Centennial Year essays to heart belonged to agricultural associations. There were not many of them in Indiana in the decade following the Centennial Celebration. Between the associations affiliated with the State Board of Agriculture, the Grange, and independent local clubs, they represented, perhaps, as many as 15 percent of Indiana's farmers. Bringing other farmers into the fold was not a priority. For their part and for good reason, many non-joining farmers wanted the organizations to focus on questions of political economy. With political action and economic cooperation alive (in their minds at least) as viable purposes, these farmers were slow to appreciate clubs that offered only a strictly agricultural education. Prominent farmers also had fresh memories of the farmers' movement. Their memories—refreshed every political season—made them wary of outreach campaigns. On the whole, they were content to mind their farms.

By the mid-1880s, when Indiana's leading men relearned the necessity of missionary work among the farmers, a new form of agricultural organization was needed. Discredited with the common farmer, neither the county agricultural societies nor the Grange could serve as forums for interchange and outreach. Inspired by the example of farmers in other states, advocates of agricultural improvement looked to a new educational institution, a series of farmer institutes (modeled after teacher institutes) hosted in each county during the winter months. Funded in part by the state legislature, and hosted jointly by public land-grant universities and leading agriculturists of the counties, the farmer institutes provided a politically neutral site for learning about practical agriculture and public policy (primarily local and state) related to it.

As an agenda of agricultural improvement, the farmer institutes' curriculum differed little from that of the agricultural societies of the 1850s. The institutes' popularity, however, testified to how much Indiana's farmers had changed in the span of two generations. In the 1850s, farmers who subscribed to the tenets of agricultural improvement were unusual farmers, men who represented the numerically small opinion of the town-oriented elite. By 1890, the farmers who subscribed to the tenets of agricultural improvement represented the prevailing opinion among Indiana's farmers.

Agricultural Improvement's Leading Edge: State Industrial Associations

Confidence in the agricultural future was concentrated among the farmers who belonged to the State Board of Agriculture and its delegate board, composed of representatives from each county. A sizable portion of the State Board's members were not farmers, or were absentee farm-owners or hobbyists; the genuine farmers attending its meetings were among the most prosperous. Since the 1850s, they had gathered annually to discuss "the wants, prospects, and conditions of the agricultural interests throughout the state" ("Act for the Encouragement of Agriculture," 1851, p. 41). Agriculturally, holding best practice discussions and making policy recommendations to the General Assembly were the State Board's main concerns. But, by the 1870s, the 2½ day annual meeting was filled with committee reports pertaining to the State Fair, with only an occasional agricultural or policy discussion (State Board of Agriculture, 1877). Notwithstanding agriculture's neglect, some town-dwelling delegates thought it received too much attention. Instead "of occupying *so much* of the time of the session in discussing culture of crops, treatment and grades of stock, etc., etc.," they thought the State Board's time would be better spent considering the "best modes" of fair management (State Board of Agriculture, 1880, p. 319). Leading farmers wanted more agricultural papers and deliberations, not fewer.

Indiana had no shortage of men who could give sensible agricultural talks. Discontented with fair discussions, they were creating specialized agricultural associations that met annually in Indianapolis. Men with an interest in importing and breeding Shorthorn cattle organized first in 1872. Others followed: poultry raisers in 1875; sheep-raisers and tile-makers in 1876; swine-breeders and dairymen in 1877; beekeepers in 1880; sorghum-growers in 1882; and more in subsequent years. Some of the members were wealthy gentlemen who amused themselves by devoting spare time and money to the agricultural arts. Most took their farming seriously; they made a particular agricultural pursuit a specialty for investigation, experimentation, and commercial exploitation. Their object in holding annual meetings was to exchange ideas in the hope of advancing the commercial prospects and earning potential of their specialty (Latta, 1938).

Although each formed separately, the specialty associations' memberships overlapped with each other and with the State Board of Agriculture. Collectively, they became known as the industrial associations because industries, such as cheese factories, grew out of commercial agriculture. By the mid-1880s, their meeting times were synchronized to be held just prior to the State Board's annual meeting in January so that each branch of agriculture could lobby for a just share of the State Fair's premium awards and offer recommendations on public policies of special concern. As livestock breeders and businessmen, advancing the interests of their chosen specialties formed the core of the industrial associations' mission.

The public benefits of the industrial associations' work came from their role in expanding the volume and diversity of agricultural produce that Indiana's farmers made available for commercial sale, as well as improving its quality. The dairy industry serves as an illustrative example. Having "run to hog too much" Hoosier farmers furnished such a "scanty supply" of butter that town-dwellers had to import it from neighboring states. The problem of scarcity was compounded by the inferior quality of the butter local farmers made available to townsfolk. Since Indiana's climate and growing conditions were well suited to dairying, her farmers' failure to furnish dairy products of good quality represented a lost opportunity for local communities to take a step toward self-sufficiency. Why should Indiana import from abroad what it could produce for itself? By advocating better cattle and improved milk-processing techniques, the Dairymen's Association could help farmers meet the growing urban demand for butter, cheese, and fluid milk. Just about every farmer could use some additional income; the dairy market was wide open to men willing to learn how to earn it (Johnson, 1881, pp. 362–363).

What applied to the dairymen applied to all of the specialty associations: Each specialty area represented an agriculturally based commercial want, and each had the potential to support manufacturing or processing industries that could move Indiana's economy toward self-sufficiency. Significantly, each industrial association was founded on a productive activity that farmers already did for home use, that is, to meet their families' consumption needs. (For instance, virtually every farm family kept a few cows; traditionally, dairying was considered to be women's work for the household.) Farmers tended to think of the business side of farming only in terms of producing a single traditional cash crop staple. They did not treat the variety of home use products as profit-making opportunities. The members of the industrial associations did. Engaging in one or two specialties alongside traditional cash crop production was how they expected farmers to make the farm pay in the future (Lockridge, 1875; Thompson, 1877).

Fundamentally, Indiana's leading agriculturists had the right idea. There was financial security in mixed husbandry, and by practicing it for home markets farmers could avoid freight charges. On the whole, though, prominent farmers were in no hurry to invite other farmers to join their agricultural discussions. A passing gesture was made in 1882, when Indiana's governor appeared before the State Board of Agriculture and State Grange to advocate cooperation and outreach among the farmers. The State Board's secretary took the message to the industrial associations, proposing a quarterly circular that would put the "sense of the meeting" on agricultural issues before the reading public (State Board of Agriculture, 1882a, p. 360). The industrial associations received the idea favorably. The next session of the General Assembly did not. It refused to subsidize printing expenses. Little else was done in the name of educational outreach. The industrial associations remained clubs for the state's wealthiest men. In the early 1890s, their combined membership was around 300 people (Clover Top, 1891).

Some industrial associations' members were determined to have an association that reached out to other farmers. Robert Mitchell assumed the Shorthorn Convention's presidency in 1885. He made clear what was at stake if the thirty-odd members continued to "slumber on" merely reading prepared papers to each other once a year. The Shorthorns were losing the "battle of the breeds" among Indiana's farmers to Jerseys (dairy) and Herefords (beef). The cattle were not at fault; good for meat and milk, the Shorthorn breed was well suited for the general farmer who supplied his family's table first, and then the market. Shorthorn men were at fault for failing to put the breed's advantages before the "average reading farmer." To prepare for an educational campaign that would make farmers aware of what breed was in their best interest, the Shorthorn men could start by castrating a few calves to give farmers an "object lesson" at the next county fair. In the meantime, while the calves fattened, they should begin forming stock-growers' associations in their home communities (Mitchell, 1885, pp. 382–384).

The "Necessity and Value of Local Effort" was made the predominant theme of the 1885 Short-horn Breeders' Convention. From it, the Short-horn men acquired a mission that was "higher, broader and more comprehensive" than simply getting good prices from the butcher. As "citizens of Indiana" its members were urged to "look to the advancement of the farmers of the State as well as [their] individual profits" (Buckles, 1885, p. 200). By all appearances, they did. A weekly Shorthorn column was launched in the *Indiana Farmer*, county stock-growers' associations took shape; a statewide directory of Shorthorn breeders was compiled; and, public auctions (to farmers, not butchers) sprang up around the state. Advice to beginners became a staple of the agricultural papers, as did testimonials from newcomers regarding how well they were received at the Shorthorn meetings (Thomas, 1890; Yorkes, 1886). Enrollment in the statewide Shorthorn Convention grew rapidly from its paltry membership of 34 in 1884 to 236 in 1888 (State Board of Agriculture, 1888). Only the best farmers attended the annual gathering, but, in contrast to a decade earlier, they represented other farmers. If they did not do so literally, as a result of election at the meetings of county livestock associations, they did so virtually and legitimately as the spokesmen for a substantial number of farmers' self-discerned interests.

The Atrophy of the Indiana State Grange

The Patrons of Husbandry showed scarcely more inclination to educate other farmers than the industrial associations. After the controversies over the 1874 election campaign and the State Purchasing Agency, leading grangers sought to put the farmers' movement at arm's length. Steering clear of partisan politics qualified as prudence. In 1870s Indiana, with memories of the Civil War still fresh, few voluntary associations could place Republicans and Democrats side-by-side and survive. Dismantling the State Purchasing Agency, however, reflected something other than prudence. That decision was a high profile expression of what was taking place in communities throughout Indiana. Grangers were turning their backs on the farmers of the countryside. The local grange was becoming an exclusive club for the more prosperous farmers who lived in and near the villages. These farmers understood well the advantages of robust home markets. Instead of activism in politics or economic cooperation, they advocated self-education in diversified commercial agriculture. This doctrine became Grange gospel.

After their experience with the 1874 state election campaign, grangers made sure that their organization did not become part of the political machinery in the 1876 presidential election campaign. The prospect of cheap paper currency and rallying farmers inspired another independent challenge, this time from the new Greenback Party. Anticipating the agitation, the State Grange refused to provide funds for the traveling lecturers who, in previous years, had carried grange teachings and enthusiasm into the countryside. Less public speaking and more farming were prescribed. Calls to reinstitute the traveling lecturer system were ignored (Advancement, 1876; Ham, 1876).

By the time the 1876 elections were safely past, the condition of many local granges was "not very flattering" (Clark, 1876, p. 5). Several leading grangers urged the State Grange to give "fostering care" to its subordinate branches (Post, 1876, p. 5). The State Grange had money enough to provide inspiring instruction, at least. At the next meeting, nearly one-third of the treasury ($3,000) was set aside to cover the traveling costs of a team of lecturers. Although the team was invited to deliver nearly 500 lectures to local granges during the next year, over half of the state fund went unused (there were roughly 2,000 local granges). Grangers, it seems, were not all that interested in having the organization's advantages explained to the other farmers of their home communities (Indiana State Grange, 1877).

In a year free of political campaigning, farmers' apathy toward strictly agricultural instruction and dissension over the State Purchasing Agency undermined the lecture system. The testimony of one traveling lecturer reveals how the two factors came together. In virtually every county he visited (he went to 32), he found "dead granges" aplenty. Some meetings had scarcely an audience. At others, the apathy toward practical agriculture talk was so profound that "a man possessed with the eloquence of a Webster, and the zeal of a Wesley, could accomplish but little." It was "a waste of time and money" to send lecturers to farmers of this sort, especially while farmers "composed of good material" were waiting in the countryside. Already "talked to death," these farmers wanted deeds, not words, and knew too well that the State Grange had a "business arm" tied behind its back. The sooner economic cooperation was re-inscribed upon Grange banners the sooner talk of the Grange would fill their hearts with "energy and zeal" (Dunn, 1878, p. 7).

No matter how much strictly agricultural education went along with it, outreach hinting of economic cooperation would not be countenanced. The State Grange refused to re-authorize the lecturing system. If local granges wanted public speakers to invigorate farmers' commitment, they would have to find and finance their own. If they wanted the Grange to keep "bearing fruit perpetually," the cultivation was "on their hands—the field at their own door" (Indiana State Grange, 1878, pp. 40–41). Most grangers were content with a garden plot inside the village limits and left the countryside's broad acres untended. The backsliding members could have filled the State Fairground and the surrendered grange charters used to wallpaper a commodious farmhouse. By 1882, only 228 local grange clubs remained in good standing (Indiana State Grange, 1882).

The attrition was so much the better, as far as many grangers were concerned. The failed local clubs were "simply the dead branches of the system" and their "lopping off was the most beneficial thing that could have happened" ("Grange Notes," 1877, p. 7). Surely, despite the waning interest of some members, there were enough solid farmers "to keep the good cause afloat" (Green, 1877, p. 7). The granges left standing were in "good trim and strong in the faith" (Lowden, 1878, p. 7). The schoolhouse meeting was a thing of the past. Grangers now enjoyed the comforts of substantial two-story halls. Discussions at monthly meetings were robust: rarely did grange-attending farmers "fail to have some interesting topic to discuss" (Banner Grange, 1878, p. 7). For entertainment, they could boast of having a "fine organ and the best choir the county affords" ("Grange Notes," 1877, p. 7). So firmly planted, so full of interest, there was "no danger of so great a calamity as the downfall of the grange." What was lost in quantity of membership was gained in quality ("Grange Notes," 1878, p. 7).

With its membership diminished considerably, the Grange now enjoyed a unity of purpose. Promoting "a spirit of brotherhood among agriculturists" and elevating agriculture "by the mental, moral and social improvement of its members" ranked high on the agenda. Well down the list—after rational tillage, strict economy in expenses, the cash system, local manufacturing, and making the home beautiful—ranked opposing "special class legislation in every form," and seeking good, honest men for public office ("True Objects," 1877, p. 7). Staffed with a "more efficient and active membership" (if far smaller), whose "whole heart and mind" was committed, the Grange would advance the farmer's interest through the mutual improvement of its members (Pencil, 1878, p. 7).

"Education" was the watchword, education "in every direction." In practical agriculture, political economy, and social refinement, farmers needed to raise themselves "up to a higher standard of manhood, to think and act independently" ("Education in the Grange," 1878, p. 7). Once they did, farmers could "make themselves heard and their influence felt" in public affairs (Dean, 1883, p. 16). Wire-pulling political party hacks and special-interest toadies would stand aside. Talk of "oppressions and encroachments on the rights of farmers by other classes" would cease. Operating at "a higher position in the scale of intelligent citizenship," the farmer would "assert and maintain his rights in every and all respects" (Jones, 1884, p. 15).

Was the turn to education a surer means than political action or economic cooperation for grangers to restore farmers to their rightful place in society? Or, was it timidity at the prospect of inadvertently stirring-up another farmer's movement? Both were involved. A few grangers kept the controversial issues alive in the State Grange, warning that "concentrated capital" was the "enemy of republican institutions," that market-cornering by corporations, rate-fixing by railroads, and backroom dealing by politicians were thriving (Indiana State Grange, 1879, pp. 8-9). Unless and until farmers became educated in agricultural and economic matters they could not expect public policy to reflect their interests. In that sense, the Grange's turn to education was the sure road to reform. However, timidity prevented action, both educative and political. Every year, the State Grange passed numerous resolutions on issues great and small. In their home communities, though, its members failed to circulate petitions or to mobilize public opinion among the farmers (Indiana State Grange, 1881, p. 9). Declaring the sense of the State Grange meeting amounted to an empty gesture. Little wonder, then, that its resolutions were ignored by lawmakers until "thrown out as waste paper" (Indiana State Grange, 1879, p. 9).

Deprived of economic cooperation and spared the "agitation of reform questions," a spirit of aimlessness prevailed among the grangers. With "no object except to kill the time," the monthly meeting degenerated "into a school of gossip," its participants scarcely more than "insipid and brainless puppets, useless alike to the world and to themselves" (Post, 1879). An enemy of the Grange could not have delivered a more scathing condemnation. That it came from State Secretary C. C. Post gave it a real bite. A well educated, prosperous farmer (definitely not a hayseed radical), Post had served as Secretary of the Indiana State Grange for five years. At the height of controversy over the State Purchasing Agency, he had urged greater attention to education. His hope, a hope shared by many, were that discussion and shared study would create agreement on what needed to be done and how to do it—member by member, grange by grange—from the ground up.

C. C. Post was disappointed. Instead of regrouping for a more methodical and sustained attack, his fellow grangers "retired from the field and slept," sated by a few crumbs tossed to them "from a table loaded down with bread and fruits stolen from" the countryside. Instead of devising a "common policy" around "co-operation and the correction of legislation through non-partizan action," grangers gathered around the organ to sing songs. Instead of forging solidarity with other farmers, they backslapped themselves for their agricultural virtue: only "good grain" remained, now that "the chaff" had blown out of the Grange. In the meantime, the Patrons of Husbandry in Indiana was "dying from lack of work" even though thousands of solid farmers were "willing to put the shoulder to the wheel." They waited only for leading grangers to "pick up the pole of the wagon and give the word of command 'forward'" (Post, 1879).

The members of the State Grange did not want energetic action to advance the interests of all farmers. Across the next four years, they did almost nothing to take the missionary spirit into the countryside. An effort to create a special lecturers' fund based upon a $2.50 donation from each subordinate grange failed. A year's wait netted $57. Delegates from only 40 counties (out of 92) attended the 1885 State Grange meeting (Indiana State Grange, 1885). The membership roll grew "small by degrees and beautifully less" from its 1876 peak of more than 60,000 members, closing out the year 1887 with around 2,000 members (Indiana State Grange, 1888, p. 17). If not quite dead, the Indiana State Grange was "palsied and dormant" (Indiana State Grange, 1887, p. 5). Too few of its members were willing to lend their mite to the cooperative self-help (educational and otherwise) that had once been hailed as the defining feature of the local grange. Busy making their home farms pay, the grangers stood by as the Patrons of Husbandry— the only farmers' organization that had ever extended its branches into the countryside—withered.

Bringing Farmers into Town for a Strictly Agricultural Education

Notwithstanding the prevailing organizational indifference to outreach, there were men who hoped to engage other farmers in agricultural education. Through the overlap of the Grange, county agricultural societies, and industrial associations, a coalition of improving farmers was taking shape. They differed in their agricultural specialties, as well as their political affiliations, but they shared common ground as practical farmers. What they lacked was a mechanism for discovering it and putting it into action. Given their history and design, none of the existing farmers' organizations could serve the purpose. The Grange's withdrawal from the broader concerns of political economy and its transmutation into a social club had embittered too many farmers; county agricultural societies had always been too preoccupied with hosting fairs. A fresh institution was needed, one free of imperatives that conflicted with the mission to educate farmers in practical agriculture and public policy related to the farming business.

This emerging coalition of farmers had learned of an innovation that offered a professional educational program for people who engaged in a calling that had never been judged worthy as a profession in the public mind. In that regard, farming shared kinship with school-teaching. Taking the county institutes held for teacher-training as their model, the farmers of Michigan and Ohio (and a few other states) began holding farmers' institutes. The practical work of the two- or three-day gatherings mixed formal lectures with discussion sessions; meals, music, and literary recitals were interspersed throughout the program. Traveling professors, professional men, and gentlemen farmers did much of the lecturing, but, as a matter of principle, every institute had an equal or greater number of people from the immediate vicinity. Following each presentation, the floor was opened for a question-and-answer session, and, once they got accustomed to it, farmers peppered the speaker with questions and comments. Sometimes inquisitive, sometimes hostile, they challenged their learned guests to put facts behind their theories, to prove that their propositions paid when put to work on the farm. By all accounts, save the occasional chronic grumbler, when they were managed well, the institute sessions were lively, entertaining, and instructive (Moss & Lass, 1988).

Why could not Indiana host its own farmer institutes? That was one idea Indiana's governor put before the State Grange and State Board of Agriculture in 1882. The state already had a network of county organizations, and, since the institute plan required the host county to pay one-half of the expenses it would be inexpensive. The Worthy Master of the State Grange thought it reasonable to ask the General Assembly for $500 to finance a trial run. The State Board's president agreed. After all, the farmers of Ohio did not "hesitate to go into the Legislature and ask for money for that purpose." Other members of the State Board had another purpose for which they wanted legislative aid. Since the failed State Fair and Exposition of 1873, it had become a biannual ritual to request an appropriation to pay the interest on the debts the State Board had assumed. Seldom did a season pass without rumors (and sometimes serious negotiations) about selling the state fairgrounds to gain relief. Although all the State Board's members agreed that $500 for farmer institutes was a "pitiful sum" in light of Indiana's "great agricultural interest," some feared that asking the General Assembly for more money would jeopardize the appropriation for interest-payments (State Board of Agriculture, 1881, pp. 90–91).

The state legislature would not convene for another year, so an advance trial of farmer institutes could hardly hurt the cause. Or could it? The State Board decided to host four institutes, two in the early months of 1882 and two in the autumn. The winter institutes seem to have been well organized, but lightly attended ("Farmers' Institute," 1882). The autumn pair suffered a worse fate. Prospective lecturers refused to respond to requests and members of agricultural societies asked to be excused from their obligations. The sessions were never held. Farmer institutes simply could not compete with the "political excitement" of the 1882 election campaign (State Board of Agriculture, 1882a, p. 58). Charged with introducing farmer institutes into Indiana, the state's leading men failed to put much effort into the project.

Determined to make the farmer institute "a permanent educational measure" in Indiana, delegates to the State Board's 1883 meeting sought a $1000 appropriation from the General Assembly (State Board of Agriculture, 1882a, p. 94). The committee sent to wait on the legislature, however, decided that farmers needed a State Fair more than they needed farmer institutes. Apprehensive that submitting the request for $1000 "might jeopardize the appropriation for the interest" on the State Fair's mortgage, the committee took it upon themselves to postpone the request until the next legislative session (State Board of Agriculture, 1883, p. 19).

Five years—not two—turned out to be how long Indiana's farmers would have to wait for public sponsorship of institutes. At the next meeting of the State Board of Agriculture's delegate board, the farmers' representatives cast a vote directing the Executive Committee to devise an operational plan for "forming Farmers' Institutes throughout the State" (State Board of Agriculture, 1886, p. 20). Instead, the Executive Committee spent its time considering the potential sale and purchase of fairgrounds. Concerned only (or mainly) with making the State Fair more successful, they hoped to move it to a new location on the outskirts of Indianapolis. Most farmers did not give a whit about the State Fair, so their delegates gave as good as they got at the next meeting of the State Board of Agriculture. Robert Mitchell's proposal to put off indefinitely the fairgrounds question "carried by a storm of yeas, with only two or three piping nays as an offset." Again, Indiana's improving farmers demanded that the State Board of Agriculture pursue public funding for institute work ("Settlement," 1887, p. 8).

The 55th session of the Indiana General Assembly turned out to be one of the least productive on record. It was deadlocked between a Republican-controlled House and a Democratic-controlled Senate. The controversy behind the deadlock (political appointments) had nothing to do with agriculture. The mere handful of bills that made it through both legislative chambers during the three-month session had nothing to do with agriculture either. Denied a public appropriation, the State Board of Agriculture placed farmer institutes on hold. The state legislature would not convene again until January of 1889 ("What the State Legislature Did," 1887).

Public funds or not, sponsorship from the State Board of Agriculture or not, leading farmers were determined that institutes would be held in 1888. Their advocates agitated the matter in their home communities and in the farm papers. As in 1882, the intent was to prove to the General Assembly that a demand for institute work existed among the farmers. Unlike the 1882 series, though, more than four demonstration institutes were planned. The *Indiana Farmer* staff, Purdue University's professors, and a number of men from the Grange, county agricultural societies, and industrial associations, were willing to volunteer time and labor. Unlike the first desultory attempt at proving the farmer institutes' merits, the second demonstration trial had vim and verve ("Farmers' Institutes," 1888).

Just how many institutes were held is unclear (20–30 is a reasonable estimate); some regions hosted several, others none. Everywhere that institutes were held, though, "large crowds" responded to the call. Checking their political persuasions at the door, farmers talked about farming, as a practical art and a business interest. The strictly agricultural discussions proved lively enough, as "eminent scientific gentlemen" discovered that "the practical farmer was 'loaded' with ideas directly in opposition to those submitted" in their papers. The controversies, "although heated at times, aroused enthusiasm," and yielded the "most profitable results." Farmers called for more meetings, and organizers received letters "from various parts of the State asking that institutes be held" in their communities (State Board of Agriculture, 1888, p. 182). The volunteers' willingness and farmers' receptiveness proved there was a robust demand for institute work.

With success and public enthusiasm behind them, the farmer institutes' champions were certain they could get legislative aid. They submitted a $5,000 appropriations request to the State Board of Agriculture. Somewhere on the way between the State Board's rooms and those of the legislature, the lobbying committee forgot its instructions. They asked for $50,000 instead, hoping to gain complete relief from the State Fair's old debts. Unknown to the lobbying committee, however, a different request for $5,000 was slipped into the legislative hopper (State Board, 1889, p. 102). The farmer institutes had friends with legislative influence. The resulting law gave the responsibility for organizing farmer institutes to Purdue University ("Act to Encourage the Study of Agriculture," 1889).

Purdue University's leadership role in Indiana's agricultural improvement was a long time in the making. In the 1850s, some members of the State Board of Agriculture had desired a publicly financed agricultural college and model farm. Although other states, particularly Ohio and Michigan, were creating agricultural schools, requests for a "liberal appropriation" from Indiana's General Assembly were ignored (State Board of Agriculture, 1857, p. 16). After the United States Congress approved the sale of public lands to support agricultural and mechanical colleges in 1862, Indiana lagged in accepting the federal largess. Determining what use of the land-grant fund was permissible, and then choosing a location for a new university, proved troublesome (X. Y. Z., 1867). After a special committee took charge, the city of Lafayette won the bidding war. Construction began, but the tight-fisted (and debt-loaded) state legislature refused to appropriate needed funds ("Our State Agricultural College," 1873). Not until 1874 did Purdue University begin to function as an institution of higher learning.

For several years, agriculture was not a priority. In what little farmers knew of Purdue's agricultural work they found ample reason to complain. Observers at a "great field trial" of farming implements in 1876 beheld a fine (if small) model farm, but they "failed to discover any attempt at experimental farming" ("Trial of Agricultural Implements," 1876, p. 4). Once initiated, the experimental work was aimed at "amateur farmers and gardeners" who grew "a few grapes on expensive trellises," not at the needs of working farmers (Goodwin, 1878, p. 4). A change in the university's management was demanded ("Purdue University," 1878).

Purdue's next president pushed the university more squarely in the scientific and practical direction. He also made a point of appearing often before the State Board of Agriculture preaching the same sermon: if it had the money and students, Purdue could do a great deal for Indiana's farmers (Smart, 1884). With the college president came chemist Harvey W. Wiley who revealed his work analyzing commercially produced fertilizers, and Charles Ingersoll, the new professor of agriculture who was creating an agricultural program and turning the farm toward useful experimentation.

Most important for the education of Indiana's farmers, though, was the arrival of William C. Latta in 1883.

A young man, and native of north-central Indiana, W. C. Latta would spend a lifetime in the service of Purdue University. Significantly, he had practical experience on his father's farm and the best education in scientific agriculture available by way of the Michigan Agricultural College. With a clear grasp of the political and popular "obstacles to agricultural education," Latta was committed to creating a system of farmer institutes (Latta, 1883, p. 207). He found many of the farmers' delegates to the State Board of Agriculture receptive to the idea, but pivotal members of the State Board proper were indifferent. If holding the State Fair on new grounds was their ambition, Professor Latta, along with most farmers, was willing to let them have it. The State Board of Agriculture got the $50,000 it wanted to pay its mortgage (and, soon thereafter, the money to buy new fairgrounds). Purdue University—supported by its newly created and federally funded experiment station—took charge of the farmers' practical agricultural education.

Acting as the Superintendent of Farmers' Institutes, Professor Latta's role was fairly circumscribed. He solicited traveling speakers, synchronized institute dates into a workable (and very cost effective) system, and handled expense-reimbursement from the fund created by the General Assembly. Everything else was left in the hands of the county institutes' chairmen and local agricultural organizations. By design, the farmer institute model relied heavily upon the voluntary participation of leading farmers. The governing principle, cooperative self-reliance, was reinforced by necessity. Indiana's annual $5,000 appropriation was less than one-half of the public money provided for institute work by the legislatures of other states. Every county made do with less than $45 in public funds to cover the costs for speakers, advertising, meals, music, and entertainment. By soliciting voluntary contributions of time, talents, and treasure from people in the townships, though, some counties husbanded their public dollars so effectively that they were able to hold several institutes in the same season (Latta, 1893).

The volunteer spirit was not in short supply. Nearly 100 traveling speakers "donated their services" in the first year (in subsequent years, modest sums were paid to defray expenses). Most of the traveling volunteers came from the agricultural press, the agricultural college and experiment station, and the industrial associations. They included "many of the most intelligent and successful farmers, fruit growers and stockmen of the State" (Smart, 1893, p. 5). They were joined by the home talent, mostly (in the early years) men of the town, the doctors, lawyers, and town businessmen who were farm owners, hobbyists, or involved in industries connected to agriculture. Inevitably, there were more "literary papers" on the programs than practical farmers desired, but the institute work—by design—was intended to be "broader in its scope" than a farmer's club. Reaching "out to men of all classes" it sought to bring local producers and consumers, farmers and businessmen, into "kindly fraternal relations" so that each would "regard the others as friendly, helpful and necessary to the highest good of all. The greatest good to the greatest number [was] the ruling purpose in the Institute work." Men who appreciated "this broad catholic creed" made the institutes thrive (Latta, 1891–1892, p. 96).

County institute chairmen were advised, in addition to banning partisan political discussion, to "recognize the mutuality of interests of town and country" (Smart, 1893, p. 5). The advice was unnecessary. The farmer institute could not function without aid from the town. For their part, townsfolk gave a hearty show of support. Railway companies granted reduced fares to traveling speakers on the institute circuit. Newspapers provided free advertising; town businesses paid the printing costs for programs and fliers (complete with advertisements, of course). Meetings were held in county courthouses and town halls. Businesses and schools closed for the occasion. The mayor was sure to give opening remarks. Town glee clubs and schoolchildren supplied music. Aspiring manufacturers and food processors came to tell farmers what types of produce they needed to meet consumer demand, as well as to explain what demands large-scale industrial processing required of particular raw farm produce. Improved varieties of seeds, along with agricultural newspapers, were distributed free of charge among the mass of farmers in attendance ("Farmers' Institutes," 1889; McMahan, 1893). Could there be a more convincing demonstration of the great esteem townsmen had for their improving farmers? Could there be a better illustration of how farmer institutes were well-calculated to bring town and country together?

Town-dwellers' enthusiastic cooperation with logistics and social features went a long way toward making farmer institutes succeed. The actual program—the driving purpose behind the institutes—was serious work intended for farmers' education. Neither the townsmen's nor the farmers' interests in public policy were neglected. Reforms in road building and public financing received prominent places. Proposed policies for livestock confinement, fencing, dog-restraint, and ditching gave farmers ample material to discuss. The practical farming topics ranged widely. Farmers heard papers on hoof rot and animal care from medical men; on commercial fertilizers from chemists; on drainage from tile-makers; and, on orchards and fruits from horticulturists. Leading farmers were placed on the program, sometimes to correct expert opinion with practical insights, other times to offer their own thoughts on making the farm pay. To farm wives, leading ladies gave advice on practicing domestic economy, making the home beautiful, and educating the family through reading circles. Discussion sessions brought out the finer points and objections. For farmers who were not satisfied with the official program, the institutes borrowed an innovation from the Grange: a query box in which people, particularly the shy sort, placed slips of paper with questions that they wished to have considered ("Farmers' 'Round-Up' Institute," 1889).

Equal parts agricultural celebration and serious education, the farmer institutes marked the first broadly based, concerted effort by leading townsmen and farmers to enlist the countryside's farmers in educational improvement. Initial reports of farmers' reactions were mixed, but the general sentiment strongly favored continuing the institute work, as well as increasing the public appropriation. After attending several sessions J. G. Kingsbury (1890) proclaimed the institute to be "unsurpassed" as "a farmer's educator" (p. 10). Coming from the man who, through nearly 20 years of editing the *Indiana Farmer*, had done more than anyone to promote agricultural education in the state, this was high praise. It was well deserved, for the number of institute-attending farmers grew steadily. By the mid-1890s, the average attendance (across all 92 counties) at daily sessions was 274, and the typical season boasted a statewide enrollment that exceeded 70,000 ("Farmers' Institutes," 1903; Latta, 1895-1896).

For men who spent the better part of their lives trying to promote agricultural education, the farmer institutes' success must have been gratifying. Holding agricultural fairs had failed to bring farmers into the serious work of the county societies; the men who held the reins wanted thriving fairs more than good farming. Farmers had joined the Grange in droves, but some had done so as a political protest, while others joined to save money. Disappointed and betrayed they dropped out. The men who remained in the Grange wanted a social club more than a working farmer's club. The institutes were different: their guiding principle was strictly agricultural education. Farmers attended because they wanted to learn how to make the farm pay, and the best farmers and scientific men were willing (if not eager) to teach them. As an institution and as an agricultural program, the farmer institute had all the necessary ingredients for teaching farmers how to adjust to—and advance—the growth of Indiana's manufacturing economy and town-dwelling population. Deservedly, it enjoyed widespread public support.

The farmer institute was not a success everywhere; nor did it appeal to all farmers. The statewide annual enrollment of 70,000 was impressive, but there were more than 200,000 farms in Indiana. In every county, there were men who knew about the annual institutes but stayed home because they were diehard opponents of book-farming. Others who were counted among the "best farmers" of the community also stayed home, because they thought "they knew all about farming and could not learn from others" (Harlan, 1893, p. 1). Breaking through the countryside's isolation also posed difficulties. Several years into the institute work, there were outlying areas where farmers remained unaware "that there was a State appropriation for that purpose." In some places, the institutes' managers complained that their efforts were "wasted on empty seats" (Latta, 1893, pp. 403–405). Sparsely attended institutes were likely to be found where the terrain was ill-suited to commercial farming or where the towns were small and far apart. There the old-style corn-and-hog farming (hilly lands of the south-central region) or the modern-style commercial farming of wheat or cattle (northern and western plains) prevailed. In counties dominated by these kinds of farming, "some slight antagonism of town and country" was evident, and farmers were likely to display an aversion to institutes (Latta, 1890, p. 8).

Most troubling to W. C. Latta, though, was that "throughout the State the *average* and *poor* farmers [had] yet to feel the energizing influence of a closer contact with their most wide-awake fellow farmers." As the Superintendent of Farmers' Institutes, he urged institute workers and prosperous farmers to get in touch with their "less successful brothers and inspire them with the laudable ambition to be first-class farmers as well as intelligent public-spirited citizens" (Latta, 1892–1893, p. 472). From an agricultural and a civic standpoint, it was the right thing to do. It was also the prudent thing to do. While the farmer institute system had been taking shape, another kind of farmers' organization had been growing among the sort of farmers who knew little of institute work. It, too, advocated education in practical agriculture and public affairs for farmers. But, the origins and ambitions of the new movement bore little resemblance to those of the farmer institute, and its public policy agenda carried a sweeping reform program that went far beyond that of the farmers' movement of the 1870s.

Reflections on Civic Learning

The decade from the mid-1870s to the mid-1880s represents the nadir of organization among Indiana's farmers. From a civic perspective this is intriguing, for opinion among the best farmers had never been more solidly in favor of agricultural improvement. At the turn of the 1880s, several models of organization—county agricultural societies, State Board of Agriculture, industrial associations, and Grange—were in operation, and

Indiana's leading farmers knew about other states' farmer institutes. They should have wanted to encourage actively other farmers' improvement. For all the grumbling about slipshod farming being a curse on the community, though, diffusing agricultural knowledge was not a priority.

Perhaps the indifference to outreach is not so remarkable. Although reform movements receive a great deal of attention from historians, tending to the farm and family was the more durable characteristic of nineteenth-century life. This was particularly true after the Civil War. Social reform campaigns of all kinds lacked the missionary zeal of the antebellum era and, with the partial exception of the temperance movement, made little headway. Voluntary associations grew, but their aims were modest and their reach circumscribed. Serving as social networks for mutual education and access to insurance, they were self-help groups for the like-minded (based, largely, on race, religion, economic class, and ethnicity). Their outlook on people's duty to the public good was not so different from that of the traditional farm family: Each group looked to its own and showed little regard for the well being of others not members. In the Gilded Age, mutual indifference was the predominant civic way.

Like-mindedness formed the basis for participation in voluntary associations. Among Indiana's farmers, differences were considerable. The contrast between Indiana's very best farmers and improvement-minded farmers was large enough to convince some that they would not be welcome in the industrial associations' meetings; on the whole, the members did little to change their minds. One step down the social hierarchy, among improvement-minded farmers, differences in attitude (on political engagement, economic cooperation, civic obligations, and relations with town-dwellers) destroyed the Grange in local communities. Farmers who remained in the Grange became convinced that there was little point in trying to bring other improving farmers into the fold. Another step down the social hierarchy, between Indiana's improvement-minded farmers and the poorest farmers who rejected entirely agricultural improvement's prescriptions, the gulf was greater still. The controversies generated by improving farmers' public policy demands were bitter and seemingly irresolvable from the 1850s to the 1880s (compulsory livestock confinement and fence maintenance, county- sponsored drainage projects and road building, and state laws to curb sheep-killing dogs, to name a few). On the fundamentals of farming, economy, and community obligations, improving farmers were at odds with their non-improving neighbors, with each other, and with the most prominent farmers of their counties. Absent a farmers' organization that possessed a coherent institutional infrastructure and effective leadership, farmers could not—or at least did not—recognize the ways in which they shared common cause.

It took time for improvement-minded farmers to perceive the value of an organizational system that was broader in scope and outlook than the typical local club of the like-minded. Based on their experiences—whether in struggling to adapt to the economy or in getting preferred policies passed—many farmers became convinced that they needed an institution devoted to the agricultural education of all farmers. To get it, improving farmers formed a loose coalition through the existing agricultural organizations. Seizing the policy-advancing initiative from the State Board of Agriculture, they put their own priorities on the public agenda and got the system of farmer institutes organized in the late 1880s. It was not a coincidence that the farmer institute provided a forum for learning about and debating local public policies related to transportation and economic development, as well as agricultural practices. The farmer institutes' method was different, but the intent was more or less the same as that of the agricultural societies of the 1850s, to convince farmers that agricultural improvement's prescriptions were in their best interest as farmers and citizens of a community that included nearby towns.

The farmer institute system was a remarkable accomplishment. It marked the first time in Indiana that a sizable contingent of farmers engaged in a collective effort on behalf of agricultural improvement's strictly educational mission. Much of the initiative for the farmer institute campaign came from actual (although prosperous) working farmers who were advancing—in a positive way—their interests as farmers. It is noteworthy that their ambitions were for the farmer without being against anything affiliated with the town. Unlike the farmers' movement of the early 1870s, these farmers were not rallying against monopolies and middlemen. Unlike the fair reform movement, they were not campaigning against the townsmen's moral degeneracy. Instead, campaigning exclusively for the farmer's education in practical agriculture and economy, champions of the institute earned widespread support. Their success marked a significant change in farmers' attitude toward self-education in their calling, as well as toward the economic progress of the local community.

The farmer institute movement is all the more remarkable because Indiana's farmers were organizing with the full support of public opinion in the towns. That said something about townsmen's confidence that, given the appropriate education, farmers would construe their self-interest in ways that would be advantageous to the local town-based economy. Their confidence was well founded. When they encountered the institute, farmers recognized that—in contrast to hosting fairs—townsmen were actually doing something for the farmer that really would be mutually beneficial to town and country. About a decade after the Centennial Year essays on the progress of agriculture, the convergence on behalf of serious agricultural education that brought the economic ambitions of townsmen and countrymen together took place. The vitality of the farmer institutes made it clear that a majority of Indiana's townsmen and farmers had learned their civics lesson.

References

Act for the encouragement of agriculture. (1851). *Indiana Laws* (35th Session), 7.

Act to encourage the study of agriculture. (1889). *Indiana Laws* (56th Session), 273–274.

Advancement. (1876). No backwards steps. *Indiana Farmer, 11*(19), 5.

Banner Grange, Cass County. (1878). *Indiana Farmer, 13*(6), 7.

Buckles, J. S. (1885). Necessity and value of local effort. Short-horn breeders' convention proceedings. In *Annual Report of the Indiana State Board of Agriculture.* Indianapolis, IN: State Board of Agriculture.

Clark, J. W. (1876). About lecturers. *Indiana Farmer, 11*(48), 5.

Clover Top. (1891). House Bill 209. *Indiana Farmer, 26*(6), 11.

Dean, W. (1883). Good from the Grange. *Indiana Farmer, 18*(16), 16.

Dunn, W. H. (1878). Grange lecturers—the business feature, etc. *Indiana Farmer, 13*(25), 7.

Education in the Grange. (1878). *Indiana Farmer, 13*(8), 7.

Farmers' institute. (1882). *Indiana Farmer, 17*(12), 4.

Farmers' institutes. (1888). *Indiana Farmer, 23*(3), 9.

Farmers' institutes. (1889). *Indiana Farmer,* 24(9), 8.

Farmers' institutes. (1903). *Yearbook of the Department of Agriculture.* Washington, DC: United States Department of Agriculture.

Farmers' 'round-up' institute. (1889). In *Annual Report of the Indiana State Board of Agriculture* (pp. 245–342). Indianapolis, IN: State Board of Agriculture.

Goodwin, T. A. (1878). Open letter to the faculty and trustees of Purdue University. *Indiana Farmer, 13*(21), 4.

Grange Notes. (1877). *Indiana Farmer, 12*(21), 7.

Grange Notes. (1878). *Indiana Farmer, 13*(8), 7.

Green, W. D. (1877). Still alive. *Indiana Farmer, 12*(23), 7.

H. (1876). Farming, east and west. *Indiana Farmer, 11*(8), 6.

Ham, B. F. (1876). From the state lecturer. *Indiana Farmer, 11*(49), 5.

Harlan, H. (1893). Farmers' institutes. *Indiana Farmer, 28*(7), 1.

Indiana State Grange. (1877). *7th Annual Proceedings.* Indianapolis, IN: Indiana State Grange.

Indiana State Grange. (1878). *8th Annual Proceedings.* Indianapolis, IN: Indiana State Grange.

Indiana State Grange. (1879). *9th Annual Proceedings.* Indianapolis, IN: Indiana State Grange.

Indiana State Grange. (1881). *11th Annual Proceedings.* Indianapolis, IN: Indiana State Grange.

Indiana State Grange. (1882). *12th Annual Proceedings.* Indianapolis, IN: Indiana State Grange.

Indiana State Grange. (1885). *15th Annual Proceedings.* Indianapolis, IN: Indiana State Grange.

Indiana State Grange. (1887). *17th Annual Proceedings.* Indianapolis, IN: Indiana State Grange.

Indiana State Grange. (1888). *18th Annual Proceedings.* Indianapolis, IN: Indiana State Grange.

J. H. (1876). Progress of American agriculture. *Indiana Farmer, 11*(8), 6.

Johnson, S. (1881). President's address. Dairymen's Association proceedings. In *Annual Report of the Indiana State Board of Agriculture.* Indianapolis, IN: State Board of Agriculture.

Jones, A. (1884). The Grange as an educator. *Indiana Farmer, 19*(4), 15.

Kingsbury, J. G. (1890). The institute the farmer's educator. *Indiana Farmer, 25*(43), 10.

Latta, W. C. (1883). The agricultural college. In *Annual Report of the Indiana State Board of Agriculture* (pp. 207–213). Indianapolis, IN: State Board of Agriculture.

Latta, W. C. (1890). Institutes. *Indiana Farmer, 25*(6), 8.

Latta, W. C. (1891–1892). Farmers' institutes, their organization and management. In *Annual Report of the Indiana State Board of Agriculture* (pp. 93–98). Indianapolis, IN: State Board of Agriculture.

Latta, W. C. (1892–1893). Report of farmers' institutes. In *Annual Report of the Indiana State Board of Agriculture* (pp. 464–472). Indianapolis, IN: State Board of Agriculture.

Latta, W. C. (1893). Farmers' institutes. In *Annual Report of the Indiana State Board of Agriculture* (pp. 401–412). Indianapolis, IN: State Board of Agriculture.

Latta, W. C. (1895–1896). Brief report of farmers' institutes. In *Annual Report of the Indiana State Board of Agriculture* (pp. 402–410). Indianapolis, IN: State Board of Agriculture.

Latta, W. C. (1938). *Outline history of Indiana agriculture.* Lafayette, Indiana: Alpha Lambda Chapter of Epsilon Sigma Phi in cooperation with Purdue University Agricultural Experiment Station and the Department of Agricultural Extension.

Lockridge, S. F. (1875). Indiana as a grazing state. Short-horn Breeders' convention. In *Annual Report of the Indiana State Board of Agriculture.* Indianapolis, IN: State Board of Agriculture.

Lowden, G. D. (1878). The Grange in Rush County. *Indiana Farmer, 13*(1), 7.

McMahan, H. F. (1893). Union County. *Indiana Farmer, 28*(7), 1.

Mitchell, R. (1885). President's address. Shorthorn Breeders' convention proceedings. In *Annual Report of the Indiana State Board of Agriculture*. Indianapolis, IN: State Board of Agriculture.

Moss, J. W., & Lass, C. B. (1988). A history of farmers' institutes. *Agricultural History, 62*(2), 150–163.

Our state agricultural college. (1873). *North Western Farmer, 8*(2), 42.

Pencil. (1878). What the Grange needs. *Indiana Farmer, 13*(6), 7.

Post, C. C. (1876). The question of lecturers. *Indiana Farmer, 11*(48), 5.

Post, C. C. (1879). To the Patrons of Husbandry in Indiana. *Indiana State Grange 9th Annual Proceedings* (unpaginated insert). Indianapolis, IN: Indiana State Grange.

Purdue University. (1878). *Indiana Farmer, 13*(52), 4.

Settlement of the fair grounds question. (1887). *Indiana Farmer, 22*(3), 8.

Smart, J. H. (1884). What can our agricultural college do for the farmers of the state? In *Annual Report of the Indiana State Board of Agriculture* (pp. 76–79). Indianapolis, IN: State Board of Agriculture.

Smart, J. H. (1893). Report on farmers' institutes conducted by Purdue University. *Indiana Farmer, 28*(6), 5.

State Board of Agriculture. (1857). Secretary's report. In *Annual Report of the Indiana State Board of Agriculture*. Indianapolis, IN: Author.

State Board of Agriculture. (1877). Proceedings. In *Annual Report of the Indiana State Board of Agriculture*. Indianapolis, IN: Author.

State Board of Agriculture. (1880). Proceedings. In *Annual Report of the Indiana State Board of Agriculture*. Indianapolis, IN: Author.

State Board of Agriculture. (1881). Proceedings. In *Annual Report of the Indiana State Board of Agriculture*. Indianapolis, IN: Author.

State Board of Agriculture. (1882a). Farmers' institutes. In *Annual Report of the Indiana State Board of Agriculture*. Indianapolis, IN: Author.

State Board of Agriculture. (1882b). Short-horn breeders' convention proceedings. In *Annual Report of the Indiana State Board of Agriculture*. Indianapolis, IN: Author.

State Board of Agriculture. (1883). Proceedings. In *Annual Report of the Indiana State Board of Agriculture*. Indianapolis, IN: Author.

State Board of Agriculture. (1885). Proceedings. In *Annual Report of the Indiana State Board of Agriculture*. Indianapolis, IN: Author.

State Board of Agriculture. (1886). Proceedings. In *Annual Report of the Indiana State Board of Agriculture*. Indianapolis, IN: Author.

State Board of Agriculture. (1888). Short-horn breeders' convention proceedings. In *Annual Report of the Indiana State Board of Agriculture*. Indianapolis, IN: Author.

State Board of Agriculture. (1889). Proceedings. In *Annual Report of the Indiana State Board of Agriculture*. Indianapolis, IN: Author.

Thomas, G. W. (1890). Advice of an experienced breeder. *Indiana Farmer, 25*(23), 5.

Thompson, J. E. (1877). Practical dairying. Dairymen's association proceedings. In *Annual Report of the Indiana State Board of Agriculture.* Indianapolis, IN: State Board of Agriculture.

Trial of agricultural implements. (1876). *Indiana Farmer, 11*(27), 4.

True objects of the Grange. (1877). *Indiana Farmer, 12*(24), 7.

What the state legislature did. (1887). *Indiana Farmer, 22*(11), 8–9.

X. Y. Z. (1867). The industrial college. *North Western Farmer, 2*(5), 84–85.

Yorkes, R. (1886). A beginner. *Indiana Farmer, 21*(9), 5.

9

Agricultural Improvement's Civic Harvest

Agrarian Populism and its Failure in Indiana

In the annals of American agriculture, the last third of the nineteenth century was a dismal time to be a farmer. Courtesy of foreign competition and the settling of the Great Plains, staple crops' value shrank steadily from the 1870s to the 1890s. Official USDA estimates for wheat and corn, for instance, put the 1890 per bushel selling price at about two-thirds of the 1870 price. The bushel of wheat that sold for one dollar in 1870 sold for 60 cents in 1890; for corn, the price comparison was 45 and 30 cents. Taken at year's end, the USDA price estimates were higher than what most farmers received for their crops. Getting 60 cents for their wheat would have pleased most Dakota farmers in 1890; the going-rate at harvest was 35 cents. Kansas farmers were offered 10 cents for their corn, a price so low that it would not cover the freight costs necessary to get it to market. Little wonder, then, that some farmers burned corn as a winter fuel. Not even poor seasons in certain regions—from hard winters, droughts, insect plagues, and disease epidemics among livestock—could restore agricultural prices to what they had once been (Goodwyn, 1976).

Civic Learning through Agricultural Improvement, pages 203–220
Copyright © 2011 by Information Age Publishing
All rights of reproduction in any form reserved.

Much of the cause for low farm profits came not from overproduction but from public policy. Congress's decision to return to the gold standard fixed the supply of circulating currency. Over time, as the overall economy and population grew, circulating media became increasingly scarce, particularly at harvest. Compelled to accept extraordinarily low prices for their crops, farmers shook their heads in disgust as the weeks passed, the demand for money slacked, and prices on the goods that they purchased returned to higher levels. Farmers could also tender a hearty vote of thanks to Congress for a full slate of protective tariffs. What farmer, after all, would fail to rejoice at hearing of the great manufacturing fortunes that were built on the high prices he paid for consumer goods, farm supplies, and implements? Some gratitude was owed to the United States Supreme Court, too, for striking down the granger laws Midwestern farmers had urged on their state legislatures in the hope of curbing the power of railroad monopolies. Given the exclusive authority to regulate interstate rail traffic, Congress failed to devise effective legislation. Rate-discrimination and price-gouging intensified. The farmer, as ever, was forced to accept the terms offered. He could not do without the railroad, even if he was fortunate enough to be located near a robust home market.

When they cut the legs out from under the farmers' movement in the mid-1870s, the grangers left a great deal of reform business unfinished. Unhindered by the threat of a farmers' boycott or cooperative system, middlemen and merchants piled on the interest and fees. Compelled to borrow to get by for a season, to pay the taxes, or to make small improvements in hope that the farm might yield profits in the next season, growing numbers of farmers placed their farms under mortgage. The number of mortgaged farms and the size of the mortgages rose steadily ("The Magazine Writers," 1890). Men who owned farms lived in fear that the next failed crop or unforeseen expense would push them into the swelling ranks of men who found employment in town-based industries. Young men found the traditional climb up the agricultural ladder from farmhand to farm-renter to farm-owner to be a long one. In the South, sharecropping and tenantry became so widespread that they became the normal condition for most farmers. What kind of independence was this for the citizens who formed the backbone of the American Republic?

Despite the worsening agricultural situation, the Indiana countryside was fairly quiet until the late 1880s. Within agricultural improvement circles, grumblers were met with the assertion that the agricultural economy had never been "on a solider basis" (Foote, 1887, p. 10). Claims of this sort were somewhat suspect since, typically, they omitted important considerations, such as labor costs, or how the dollar's changing value affected living expenses. Specious reasoning aside, three revealing qualifiers were attached. The first qualifier was that farmers who were free and clear of a mortgage (as distinct from debtors) enjoyed reasonable prosperity from the farm's commercial and subsistence sectors. The second was that farming paid "equal to any other safe business, when the same amount of care, labor and intelligence [were] applied to it" ("Washington County Report," 1882, p. 301). The third qualifier was the necessity of "mixed husbandry," engaging in livestock and crop production of various kinds to meet the farm family's needs and to meet the demands of urban markets near and distant. Taken together, the three criteria added up to a single conclusion: "The thinking and reading farmer [was] not troubled or perplexed in keeping even with the condition of things that relate to his progress" ("Farmers and Businessmen," 1886, p. 8).

Farmers who met all three tests—and their number included practical farmers—were "sick and tired of this cry that farming don't pay." It was true that there were plenty of farmers who could not make the farm pay, but "in most cases it [was] their fault and not the fault of the business. Intelligence is the basis of any success, and the right kind is needed on the farm as much as it is needed anywhere in this world." Farmers who grumbled and croaked about the great difficulties involved in making a living from farming were "only advertising to the world their own ignorance and incapability." For their part, "intelligent farmers" did not want to be lumped into the same category (Farmer, 1887, p. 4).

Men who did think that farmers had legitimate grounds for complaining about the state of the agricultural economy urged other farmers to spend more time reading agricultural papers and thinking about how to apply business principles to their farm management. Minding the farm was not their only prescription, however. They expected farmers to draw political lessons from their agricultural studies and to "cast their votes in the right direction" (Bismark, 1878, p. 7). The exact direction in which votes were to be directed—in terms of partisan affiliation—was rarely indicated in letters sent to the agricultural papers. That was precisely the point. If farmers took agricultural improvement seriously, they would realize that their genuine economic interests were not affiliated with any political party ("Politics in Agricultural Papers," 1884, p. 7).

Most farmers, by contrast, knew only the party spirit. They ran "wild with political excitement," parading with torches, shouting "hurrah" for their party's candidates, and glorifying office-seekers who paid appropriate homage to the farmer and the people. Seldom did farmers "stop to think" about whether their "highest interests" as farmers were being served by the men they voted into office (Stahl, 1884, p. 1). The results of farmers' unthinking partisanship were obvious. In the 1880s, farmers represented a majority of the population in over one-half of the states and three-fourths of the nation's congressional districts. Yet, of the 325 members of the United States House of Representatives, only 18 were farmers. The proportion of farmer-delegates in the typical state legislature was higher, but not sufficiently so ("Who Sent Them There," 1884, p. 9). Was it any wonder that "every other class" got a pull at the "political taffy" while the farmer's interest was ignored by lawmakers (Stahl, 1884, p. 1)?

The charge was not for farmers to create "class organizations for political purposes." Instead, the farmer was instructed to go to the primaries and polls as an "independent citizen," not as man "blinded by party zeal," to "look at the man and not the party" and to put "his vote where it will do the most good" (Stahl, 1884, p. 1). This sensible prescription was easier to preach than to practice. Grangers had always pledged good government, non-partisanship, and the election of good, honest men to office as part of their creed. Yet, in 1883, after Indiana's Democratic-controlled General Assembly treated their resolutions on various public policies with "profound contempt" by ignoring them, the grangers showed the feeble strength of their convictions. A two-part resolution was placed before the next State Grange. The first part declared it to be the duty of all grangers to work to nominate only candidates who adhered to the principle, "the greatest good to the greatest number." Those efforts failing, the second part declared that grangers would "lay down party prejudice" and cross party lines with their "ballots and influence" to support the right kind of men for public office. The convention ratified the resolution only after the second part was deleted (Indiana State Grange, 1883, pp. 20–21, 29–30). Since the grangers continued to place political party loyalties above the principles of their Declaration of Purposes, there was little reason for the next legislative session, or the following one, to act on Grange resolutions.

Indiana's state legislators could defy the Grange with impunity. With only several thousand members, the State Grange did not speak on behalf of the roughly 200,000 farm families of Indiana. The same was true in other Midwestern states. Farmers were not organized, even though every other major economic interest was. Manufacturers kept the tariff question on the congressional agenda; the railroad companies greased the political machinery to block regulation; and boards of trade, loan associations, and mercantile associations appointed full-time lobbyists to ensure that their legislative interests were protected (Halstead, 1886). Was it not long overdue that farmers should look after their own interests in public policy? Slowly, as the farm mortgages grew, the evidence of price fixing in the commodities markets multiplied, and the collusion of corporations squeezed the profits out of the farm, the answer from the countryside emerged in the form of a second farmer's movement demanding fundamental reforms in the economic and political system (Trusler, 1888).

The farmers' movement of the early 1870s had failed to achieve its economic and political reform goals, but it broadcast three seeds throughout the American countryside: organization, cooperation, and education. As the Grange atrophied, these seeds—spread far and wide by a nation of farmers on the move—germinated and took root. New farmers' organizations sprouted. In the northern plains, an alliance of clubs formed around the nucleus provided by the Chicago farm paper, *The Western Rural.* The alliance existed mainly on paper, but farmers in the Dakotas and Kansas gave it real life in the mid-1880s when they put cooperation to work against implement dealers and stockyard owners. Cooperation against local middlemen gave rise to the Farmers' Mutual Benefit Association (FMBA) in southern Illinois and southern Indiana. In Louisiana, a Farmers' Union was established, and in North Carolina, a Farmers' Association. Clubs affiliated with the Agricultural Wheel spread throughout Arkansas. In the South, the furnishings man (merchant and/or landowner) gave farmers ample reason to band together to purchase farm supplies and groceries; a cotton-bagging trust gave them another. Northern wheat farmers, in turn, got similar inspiration from local middlemen and merchants, as well as a binder twine trust. It took time for farmers' confidence in organization to rebuild, but by the late 1880s, farmers throughout the South and West were resolving to cooperate together to protect their economic interests (Goodwyn, 1976; Hicks, 1931).

The new organizations took much of their doctrine and institutional apparatus directly from the Grange. In contrast to the early Grange, though, membership was restricted to farmers only. The most important of the new organizations came from the cotton lands of Texas. There the trappings of the Grange were turned into a Farmers' Alliance that had ambitious heads in the leadership and cooperative teeth. Plagued initially by partisan politics, the Texas Alliance threw its energies into creating a system of trade stores and a statewide exchange for handling cotton sales and purchasing goods and groceries. Its leadership did not forget that public policy was behind much of the abuse inflicted upon the farmer; nor did it forget that speaking with a unified voice in public affairs could help the farmer correct the situation. The 1887 meeting of the Texas State Alliance in Cleburne signaled an end to the era of silent suffering among farmers. Instead of the usual resolutions requesting public action, the convention drew up a slate of demands that left no doubt that a sweeping overhaul of monetary, regulatory, and land policy was in order. The aims of organizing farmers and educating legislators closed their platform (McMath, 1993).

The Texas Alliance meant business. Organizers canvassed the state carrying the Cleburne demands and an armful of instructional documents. A year later, they blanketed every state of the old Confederacy in "the most massive organizing drive by any citizen institution of nineteenth-century America." Everywhere the Texas-based organizers found the farmers "like unto ripe fruit": with a "gentle shake of the bush" they brought in the harvest (Goodwyn, 1976, pp. 88, 92). A series of conferences, negotiations, and mergers among the leaders of the various southern farmers' groups accompanied the membership campaign. By the close of 1889, membership in the Farmers' Alliance of the southern states was approaching one million farmers, not counting African Americans who were restricted to a Colored Farmers' Alliance.

Throughout 1889, southern Alliance organizers were making forays into the northern states and its leaders were looking to consolidate with the northern farmers' organizations. A meeting of the organizations—principally, the southern Alliance, northern Alliance, and FMBA—was called for December in St. Louis. A cross-sectional merger proved impossible, but delegates generally agreed on the provisions that emerged in the St. Louis Platform. Economic cooperation was already encountering the same difficulties of the earlier granger movement. The updated slate of demands made it clear that the farmers were looking for political solutions. Among those that enjoyed widespread support (if not immediate ratification in St. Louis) were demands for short term, low interest, federal government loans to farmers (based on land or anticipated crop values); abolition of the national banking system and the adoption of greenbacks as legal tender; free and unlimited coinage of silver; government ownership of the railroads; equal property taxation and a graduated income tax; an end to speculation in agricultural commodities futures; and strict economy in governmental expenses (McMath, 1993).

With slight modifications—and the addition of a rhetorical masterpiece of a preamble—the provisions of the St. Louis platform would become the Omaha platform adopted by the Populist Party in 1892. In 1890, though, the farmers' organizations were not intent on creating a new political party. They were determined to educate farmers in their economic interests and to elect good men to office, not as Alliance or farmer party candidates, but as men who would honor the farmers' grievances with concrete legislative action. Justice to all men (sometimes rendered "equal and exact justice to all, with special privileges and immunities to none" in the old Jacksonian spirit, especially among the FMBA) was their "prime motive"; the "greatest good to the greatest number" the end sought; and, "education and the ballot" the means for attaining it (Cory, 1890a, p. 1).

With their political agenda solidified, members of the farmers' organizations redoubled their outreach efforts in 1890. By late spring, the Indiana countryside was well organized. FMBA organizers had been at work in Indiana for two years already; in March, delegates representing 22 southwestern counties and more than 21,000 farmers convened for their State Assembly (J. H., 1890). Organizers from the two farmers' alliances followed suit with state conventions in May and June, each representing several thousand farmers from various parts of the state ("State Farmers' Alliance," 1890; "State Alliance Meeting," 1890). The new organizations took shape alongside and through the preexisting infrastructure created by the older county societies, granges, livestock associations, and the farmer institutes. All were convinced that it was time for the farmers to eschew partisanship and work together to advance their economic interests. Even the President of the Wool-Growers' Association thought it was so. To his staunchly conservative audience of wealthy farmers he posed the same question as farmer-activists of more radical temperament: "Is it not time that the farmers organize and take this government into their own hands" (Cotton, 1889, p. 503)?

Organization, cooperation, and education were the watchwords. There was no doubt that farmers intended to put them to political purposes in the election campaign of 1890. Major changes in American political economy were in order; on that, all farmers could agree. But, would farmers work through the existing political parties to select candidates who would advance their interests (as established at St. Louis) or would they "revolt" by creating an independent political party? In early June, farmers and their allies in South Dakota and Kansas chose revolt. South Dakotans launched an Independent Party; their counterparts in Kansas joined hands with industrial workingmen to create the People's Party. Their actions sent shockwaves through the political establishments of their respective states. Their example inspired a clamor to do likewise in other states (Goodwyn, 1976, p. 176).

Indiana's farmers also held a June convention. The "largest gathering of farmers ever held in Indianapolis" (more than 200 delegates) furnished ample testimony that the farmers were going to be a "power in politics." The several farmers' organizations agreed to form a State Farmers' League to advance the farmer's interest in the election campaign and subsequent legislative session. They passed resolutions that corresponded to those passed by farmers in other states, and they insisted that the "leading political parties" nominate "men in sympathy" with their demands. A "hot debate ensued" when some delegates tried to insist that, to be acceptable, nominees "must be educated to the farmers' interest either by education or occupation." No "sentiment expressed by speakers," however, "was more often or warmly cheered than of stepping over party lines, if necessary" to secure the right kind of men for public office. If the political parties failed to nominate men who supported farmers' interests, the delegates pledged "to nominate such men independently, and strive by all honorable means to secure their election" ("The Farmers' Convention," 1890, pp. 8–9).

A farmers' revolt was not on the agenda in Indiana. After the June convention, no county-level farmers' leagues were created to coordinate the work of the organizations in their home communities (Cory, 1890b; Hamilton, 1890). Many farmers belonged to more than one local farmers' association, anyway. As members of their respective political parties, farmers pushed for and obtained a host of candidates who were in sympathy with their interests. Nominations of the major political parties—including nominees for the secretary of state position, the highest state office up for grabs—were filled with farmers. On the whole, members of the farmers' organizations seemed to have been satisfied with the men selected. Virtually no agitation for an independent party appeared in the pages of the *Indiana Farmer* during the summer, even though editor-owner J. G. Kingsbury had joined the Alliance.

In late September, though, people who were not pleased with the men chosen by the main political parties of their counties gathered to organize an independent People's Party for the State of Indiana. In the 1890 elections, it received only a few thousand more votes than the Prohibitionist Party, the single-issue party that had replaced the Greenback Party as the perennial campaign sideshow. In subsequent years, the Populist Party fared little better, despite the onset of a major economic depression in 1893 and the selection of Charles Robinson, FMBA president, as its gubernatorial candidate. With Robinson at the head, the Populist ticket received less than 5 percent of the popular vote (Phillips, 1968). It garnered less than the Greenback Party earned at its peak in 1878, and had far less impact upon Indiana's political landscape than the farmers' movement of 1874. Waged on farmers' behalf, the Populist campaign failed to generate much enthusiasm across the Indiana countryside.

Home Markets, Civism, and Learning in Indiana

At first blush, the failure of agrarian populism in Indiana may seem odd. In the early 1890s, Indiana was decidedly an agricultural state. More than 60 percent of its population lived on farms, and, like their counterparts in the South and the West, Hoosier farmers had good reasons for feeling betrayed by the American political and economic system. Nevertheless, they did not throw their weight behind the kind of political revolt that took place in Kansas, South Dakota, and several other states. For all the talk about exercising independent judgment, Indiana's farmers remained true to the party in the state elections of 1890. They gave a sweeping victory to the Democratic Party. Yet, it should not be assumed that they did so out of blind party allegiance, or out of disregard for their economic interests as farmers.

With a political culture that took shape simultaneously with the rise of the second American political party system in the 1830s, partisan loyalties in Indiana had always been strong. Men thought of themselves as Democrats or Republicans even if they understood little about policy particulars. The parties' general principles (limited, local, and libertarian versus active, county/state, and paternalistic governance) were sufficient to tell farmers what each represented. Party principle was reinforced by personal traits inherited from one's family. Ethnicity and religion were markers of political affiliation in Indiana as elsewhere, but place of family origin was equally, if not more, important. People descended from northern stock tended to be Republicans, and people from southern stock, Democrats. The Civil War and partisan demagoguery afterwards hardened the sectional alignment. The policy agenda of the Republican Party contributed greatly as well. Its advocacy of the protective tariff (manufacturing industries) made it the choice of men of the towns. The net result was that, to a considerable degree, in the popular mind, the Republican Party was the party of the town while the Democratic Party was the party of the farmer.

Yet, the Republican Party also claimed to be the farmer's party. From the outset, it boasted the smaller number of farmers, but, undeniably, it attracted the greater proportion of the best farmers (who lived close to town). Those who joined the Republican Party earliest could be found most readily in the central and northern parts of Indiana. The allegiance of some stemmed from birthright, anti-slavery, and Union credentials; for others, allegiance stemmed from commitment to temperance and other moral principles. The ultimate drawing card for farmers though, if reading agricultural newspapers taught them anything, was economic development, specifically the growth of home manufacturing and markets.

The Republican Party's industry- and town-oriented agenda is well known, but it had an agricultural counterpart, primarily in the sense that economic development could not advance without a broadly based agricultural foundation. In tandem with this went a host of local policy issues that were of concern to improving farmers but have been largely forgotten with the passage of time: laws related to fencing, livestock confinement, dog-restraint (to prevent sheep-killing), sanitary commissions to check the spread of infectious disease among livestock (caused by importations), cooperative and publicly funded drainage projects, and public road improvement. Beginning in the 1850s, these kinds of local improvement policies were advocated by leading farmers and townsmen. They were opposed, or at least, resented by farmers who were less improvement-minded. With both groups lobbying Indiana's lawmakers, the state's policies moved back and forth repeatedly between the 1850s and the 1880s. Enacted by the Republican Party when it gained control of the Indiana General Assembly, the local improvement policy agenda was overturned or watered-down by successive Democratic-controlled legislatures, until the mid-1880s, when partisan deadlock prevailed.

The second farmers' movement broke the partisan stalemate. The initial signs of change, however, came before the Farmers' Alliance campaign made substantial inroads into Indiana in 1890. Indiana's farmers had already been organizing. The campaign by the Shorthorn Breeders' Association to promote improved cattle among the farmers (as part of the mixed husbandry package) had been underway since 1885. Agitation for farmer institutes picked up about the same time, and the series of demonstration institutes was held in 1888. These organizing efforts were intended to promote agricultural improvement, to instruct farmers in their true interests, and to enlist them in a broad coalition to ensure that the interests of agriculture were represented in public policy.

The several strands of the agricultural improvement campaign were not political in the partisan sense. They were intended to have an effect on the political system, though, and they achieved the desired effect. The 56th session of the General Assembly (1889) was the most bipartisan legislative session since the early 1850s. Its famous achievement was election reform, with the introduction of the secret ballot (the Indiana ballot, soon copied by other states, listed offices and candidates in columns by party affiliation). Less famously, the 56th General Assembly created the livestock commission that improving farmers had been demanding for a decade. It gave its first generous appropriation to Purdue University for agricultural purposes and granted the farmer institute system its first public appropriations. (It also agreed to pay off the State Fair's debts.) These things—from electoral reform to public financing for new state institutions to educational outreach—have been stamped by the history books as Republican causes. Yet, these measures were passed by a state legislature that was controlled by the Democratic Party (the governor was Republican). Agricultural improvers had always claimed that their notions of what was in the best interest of agriculture were above party. The 56th session of the General Assembly marked the first time that public opinion in Indiana agreed with them (Phillips, 1968).

The farmers' 1890 campaign to put good men in public office gave the Democratic Party an even more commanding hold over the 57th session of the General Assembly. In contrast to earlier years, the agricultural improvement agenda suffered no setbacks or reversals of policy. The Indiana state legislature gave improving farmers everything they wanted, with the exception of farmers' demand for reducing the salaries of county officials. (Influenced by a ring of county officers, the resulting legislation was a convoluted farce.) An attempt was made to reorganize the State Board of Agriculture under direct state authority and the control of working farmers. That law was nullified by the State Supreme Court, but everything else stood: revision to ditching and drainage laws; changes in road building and maintenance; a strong dog law to protect sheep (licensing, taxation, and restraint); revisions to livestock roaming and fencing policies; an "equalization" of taxation on corporate property, and a $75,000 appropriation to create an Indiana display for the upcoming Chicago's World Fair (the 1893 Columbian Exhibition). Since the 1850s, leading farmers and townsmen had been pushing for these kinds of policies. Not until the 1880s did a decisive majority of farmers—speaking indirectly through their farmer organizations and the election polls—conclude that their interests as farmers were identical to what had long been hailed as the public good of agricultural improvement ("The New Laws," 1891).

At the risk of sounding a note of economic determinism, it seems reasonable to conclude that the growing presence of home markets altered the tenor of the second farmers' movement when it reached Indiana. Indiana's urban centers were not booming in the 1880s. But, where urban growth, railroad lines, manufacturing, and prime farming acreage converged, the economy was growing. Farmers who lived in the central region felt it most keenly. Access to Logansport, a city of 15,000 inhabitants, and nine rail lines running through the vicinity, gave Cass County farmers "more than a living" ("Cass County Report," 1883, p. 253) Their new farm houses, fine two-story structures made of brick, stone, or lumber, would have been a credit to any town or city. Livestock were sheltered in commodious barns; the old-style Virginia worm rail fences were disappearing fast; and, the best horse-drawn machinery could be seen in the fields. Farms throughout the county scarcely resembled their appearance of ten years earlier ("Cass County Report," 1884). The city of Muncie was only about half the size of Logansport. Nevertheless, with "the largest flax bagging and tow machinery mills in the world," two large flouring mills, and several other facilities, Muncie's surrounding county could claim to be a "manufacturing county" as well as "the Eden of Indiana for farms" ("Delaware County Report," 1877, p. 178; 1886, p. 236). Johnson County possessed no sizable cities, but its farms were in a "high state of cultivation." Two large starch works took the corn off farmers' hands; a dozen flouring mills took care of the wheat; and, farmers found ready markets for all kinds of produce in a handful of small villages and nearby Indianapolis. All told, farmers situated so favorably had "no great reason to complain" ("Johnson County Report," 1883, p. 276).

The reports of farmers' prosperity came from secretaries of county agricultural societies (usually not farmers) who, inevitably, were always looking for signs of progress to extol. Nevertheless, their reports contained ingredients of truth. For all their grumbling about hard times (and times were hard for many farmers), Indiana's farmers were better off than their counterparts West and South. To a far greater degree, Indiana's farmers practiced some variation of mixed husbandry. They met most of their family's consumption needs from what could be produced on their home farms, rather than purchasing essential groceries, and, with towns nearby, they were more likely to earn income from selling surplus foodstuffs (butter, eggs, and vegetables). Indiana's farmers were also sheltered from the worst of the evils afflicting farmers elsewhere. It cost them less to transport crops to market, and they earned higher profits on staple products (wheat, corn, beeves, and hogs) because nearby towns offered slightly better prices than the Chicago exchange. Somewhat ironically, participation in the broader Alliance-Populist movement may have made Indiana's farmers more aware of their comparative advantages and the extent to which those advantages were dependent on the local town-based economy.

Hoosier farmers' advantages had a great deal to with Indiana's built environment of transportation infrastructure, small cities, and small farms. They were results of the home market that agricultural improvers had been trying to grow since the 1850s. The home market that existed in Indiana during the 1880s was small in scale compared to what it would become in the twentieth century, but it was geographically broad. Extending outward from Indianapolis and following major transportation routes, the emerging home market encompassed roughly two-thirds of the state's area. In the southernmost and westernmost parts of Indiana—in places where towns were dispersed and industries scarce—farmers found themselves burdened with mortgage payments and high-interest loans (seasonal loans on goods and groceries). In those areas, agrarian populism flourished, but it held little appeal for farmers who resided in places that had made progress in developing the manufacturing sector of the town economy.

In light of the influences of home markets and agricultural improvement, one might suggest that agrarian populism's failure in Indiana reveals that farmers had learned three things: political engagement (in addition to minding the farm) was a necessary part of their job description as farmers; their interests as farmers required public policies that favored improvement and an enlarged role for government; and, their interests as farmers were advanced best in ways that supported the economic growth of local towns. In sharp contrast to the farmers' movement of the 1870s, the leaders of the second farmers' movement in Indiana had little difficulty convincing their farmer-followers to subscribe to a common set of goals. Nor, did they find it necessary to subvert the expressed will of their own organizations in order to safeguard the economic well being of nearby towns. To the extent that Indiana's farmers raised more hell and less corn in agrarian populist fashion, they shook their fists in fury at the great national corporations and national government. Those were in distant cities, not the nearby county seat. The farmer organizations' reform agenda posed no threat to Indiana's towns.

How much of the contrast between the first and second farmers' movements in Indiana can be attributed to townsmen's learning about farmers' situation (and sympathizing with it)? How much should be attributed to farmers' learning from experience about the advantages of mixed husbandry for home markets? How much to the agricultural and political education farmers received from their voluntary associations? Precise answers to these questions cannot be disentangled. Agricultural improvement—from first to last—was a civic agenda that emanated from towns into the countryside and from the top of rural society downwards. Perhaps some learning flowed back into town and up the social hierarchy? Changes made to the county fairs in the 1880s and the creation of county institutes for farmers suggests that some did. By reforming fairs and hosting farmer institutes, townsmen provided solid evidence that they were listening to improvement-minded farmers and that they took farmers' concerns seriously.

Unquestionably, the slogan, greatest good to the greatest number, went through some serious contortions of interpretation. In the 1850s, Governor Joseph A. Wright and (some of) the men who launched agricultural improvement in Indiana had preached it with an emphasis on dispersion, that is, on the greatest number receiving the benefits of economic development. In the 1870s, county fair organizers extolled the greatest good to justify their actions and showed little regard to whether people other than themselves benefitted. Members of the industrial associations and Grange, more or less, adopted the same attitude. They could not be accused of harming public morality, as fair organizers could, but they did little to advance the economic and educational well being of other farmers. By the mid-1880s, the greatest good to the greatest number had returned to something resembling its earlier meaning. Sounded by farmer-spokesmen like Professor W. C. Latta, Robert Mitchell, and J. G. Kingsbury, it was put to work on farmers' behalf. Education, improvement policies, and moderate reform were stressed, to be sure, but it was a marked change from the 1870s. In the late 1880s, the men who took the lead in advocating agricultural improvement took their work on behalf of other farmers seriously.

The question of how much farmers learned from experience with home markets is harder to answer. It took the better part of two generations before Indiana's town-dwelling population and manufacturing industries reached the critical mass necessary to provide a robust home market for a substantial portion of the state's farmers. In the meantime, roughly speaking, farmers' learning about the home market's benefits diminished with distance into the countryside. The presence of improved roads and a substantial urban population in a county extended the reach of the home market's educational force. While the home marketing potential grew in the 1870s and 1880s, the proportion of Indiana's steadfast opponents of book-farming and agricultural improvement diminished. Facing the imperative to improve or move, some farmers chose to move to more westerly regions. Among those who stayed behind, a younger set of farmers came of age, men who, as boys, were exposed to agricultural improvement literature, attended agricultural fairs, and identified successful local farmers as their models for emulation.

Undoubtedly, such men acquired ideas about what good farming entailed that were somewhat different from their fathers' ideas. Above all, if a farm boy came of age in the 1870s, he would have learned—courtesy of low staple crop prices—that the normal state of agricultural economy demanded mixed husbandry. For him, diversified commercial farming was not only the right thing for a good farmer to do, but necessary if he wanted to be a farmer who lived in Indiana. A man with this kind of experience would have seized eagerly the agricultural, economic, and political education offered by farmers' organizations in the late 1880s. When he did, he would have been certain that he was advancing his true interests as a farmer along with the good of his surrounding county. After all, the agricultural organizations enjoyed the blessing, the participation, and the leadership of the best men across town and country in the rural communities.

Reflections on Civic Learning

Through the contrast with the experience of the West and South, agrarian populism's failure in Indiana reveals how much agricultural improvement and home markets had made Indiana's farmers civic-minded. To make this claim is not to assert that farmers thought of themselves as townsmen or that they were in full accord with town-dwellers on all issues. Rather, it is to assert that Hoosier farmers, on the whole, had come to accept the inevitability of economic development and its implications for their lot in life. Men who wanted to farm had to alter their farming behavior to fit the opportunities offered by the town-based home market. Self-education in agriculture and public policy related to it was the way to do it. Farmers' self-interest demanded this course of action, but it was a brand of self-interest that supported the well being of the broader community. From a historical distance, this change in Hoosier farmers' outlook does not seem all that significant. In the context of the late nineteenth century, this was precisely the type of civic-mindedness that was desired.

With these things in mind, agricultural improvement might be thought of as a program of agricultural civic education. One of civic education's primary purposes is always the aim of equipping individuals to make sense of, and to some extent conform to, the world about them. Characteristically, idealizations (if inchoate at times) of the good person and the good society are involved in civism; characteristically, those idealizations reflect the considered opinion of respectable society. Agricultural improvement clearly had its image of the good producer fixed in its good farmer. Just what kind of person took shape around the good farmer (in the sense of social relations) was usually left unspecified, but the word portraits of slipshod farmers in the agricultural press leave no doubt that the bad farmer neglected his family's well-being (by loafing at the corner store, drinking, or slovenly living) and was a deadweight, if not a curse, to the community by his land-skinning, single-crop cultivation (Pencil, 1878; Sneezer, 1887; J. S. H., 1887).

Similarly, with respect to the good society, agricultural improvement's call for progress and a home market usually focused on the economic side of things, not the social relations between townsmen and farmers. An elevated level of respect paid to farmers by townspeople was certainly a part of the agricultural improvement package. Farmers' commentaries about how townsmen treated and judged farmers (particularly their children) appeared regularly in the agricultural papers. These, however, were usually placed alongside short stories, gardening, and home economy in the section tailored for the ladies among the readership. Typically, agricultural improvers did not preach about the camaraderie, sense of belonging, and pleasures of participation in the town's social life that farmers might enjoy if their prescriptions were followed. Instead, agricultural improvers homed in on farmers' economic self-interest and yoked that to the good of the broader community.

There were good reasons for this attention to economics and neglect of matters social. Farmers' primary relationship to the town was an economic one. They lived in the country, some distance from town (and county roads in Indiana were horrendous, even by nineteenth-century standards, until the 1880s in the most progressive counties). The good life for farmers was to be found on their farms with their families, a good life supplemented by the material comforts and trappings of industrial civilization. Farmers' attachment to place was through their personal investment (time, labor, and money) in their farms. This attachment was not very sentimental to be sure, but it made possible the good life on the farm (which could have its sentimental side) and pulled the farmer more closely into the orbit of town.

The farmer's attachment to place was something that agricultural improvers encouraged. Properly understood, the little farm well tilled was a lifetime project and agricultural improvers stuck to the doctrine that one move was enough (or perhaps two). The good farmer-citizen stayed put. By improving his farm he made his contribution to the good of the surrounding community. To the extent that he was expected to engage in politics, he was expected to advocate public policies that furthered his interests as a farmer. Those agricultural policies, if the farmer understood his business properly, would be advantageous to the nearby town's development. For people living in late nineteenth-century Indiana, achieving this degree of civic attachment among farmers was just the right dose.

The agenda of nineteenth-century agricultural improvement may not seem very civic when judged by the standards of a later age. We (or at least the civic educators among us) expect much more than economic self-interest to be involved. We expect sentimental attachments (patriotism and pride in our communities), comprehension of the principles of constitutionalism, and a robust concern for the well being of others. People in rural Indiana had similar concerns, but they were carried out, in the main, through agencies other than the institutions of agricultural improvement. Those kinds of aims were advanced in families' education, churches, schools, political rallies, mutual aid and charitable organizations, community events, and, simply, in people's expectations of a good neighbor's behavior. Agricultural improvement's education overlapped with the work of these educative forces, but its primary mission was to teach the farmer how his self-interest was best served by farming in ways that benefitted the surrounding community.

Without question, agricultural improvement's promoters acted with more regard for their own self-interest than for the benefit of other farmers, particularly among the county fair societies of the 1870s, the granges after the farmers' movement, and the industrial associations. But they did so in the name of the public good, and in accord with what the respectable opinion of the age deemed appropriate behavior. Perhaps, unintentionally, a steady diet of this combination of self-interest and public justification contributed to other farmers' civic learning? In any event, if the agenda of agricultural improvement was to harness the power of self-interest to pull farmers into the broader economy, it worked. If it was to convince farmers that they ought to take an interest in the progress and governance of their local communities, it worked. By the late 1880s, Indiana's farmers (at least those in organizations) took engagement in agricultural improvement and politics seriously. They had become the kind of farmers—the solid, conservative citizens—that respectable opinion had been insisting they ought to be since the 1850s. As their state's fledgling manufacturing industries matured and flourished in the early decades of the twentieth century, Indiana's farmers enjoyed the full benefits of country life in communities that had brought the loom, anvil, and plow closer together.

References

Bismark. (1878). Hamilton County Grange. *Indiana Farmer, 13*(8), 7.

Cass County report. (1883). In *Annual Report of the Indiana State Board of Agriculture*. Indianapolis, IN: State Board of Agriculture.

Cass County report. (1884). In *Annual Report of the Indiana State Board of Agriculture*. Indianapolis, IN: State Board of Agriculture.

Cory, W. (1890a). A word as to the Alliance. *Indiana Farmer, 25*(13), 1.

Cory, W. (1890b). Time to take definite action. *Indiana Farmer, 25*(38), 13.

Cotton, I. N. (1889). President's address, Wool-growers' Association. In *Annual Report of the Indiana State Board of Agriculture*. Indianapolis, IN: State Board of Agriculture.

Delaware County report. (1877). In *Annual Report of the Indiana State Board of Agriculture*. Indianapolis, IN: State Board of Agriculture.

Delaware County report. (1886). In *Annual Report of the Indiana State Board of Agriculture*. Indianapolis, IN: State Board of Agriculture.

Farmer. (1887). Farming does pay. *Indiana Farmer, 22*(23), 4.

Farmers and businessmen. (1886). *Indiana Farmer, 21*(50), 8.

The farmers' convention of the 19th. (1890). *Indiana Farmer, 25*(27), 8–9.

Foote, J. A. (1887). Were former times better than now? *Indiana Farmer, 22*(10), 10.

Goodwyn, L. (1976). *Democratic promise: The Populist moment in America.* New York: Oxford University Press.

Halstead, R. (1886). Farmers and organization. *Indiana Farmer, 21*(4), 1.

Hamilton, W. (1890). Let county leagues be organized. *Indiana Farmer, 25*(29), 5.

Hicks, J. D. (1931). *The Populist revolt: A history of the Farmers' Alliance and the People's Party.* Minneapolis: University of Minnesota Press.

Indiana State Grange. (1883). *13th Annual Proceedings*. Indianapolis, IN: Author.

J. H. (1890). Farmers Mutual Benefit Association. *Indiana Farmer, 25* (9), 14.

Johnson County report. (1883). In *Annual Report of the Indiana State Board of Agriculture*. Indianapolis, IN: State Board of Agriculture.

J. S. H. (1887). Down to Uncle Jim's. *Indiana Farmer, 22*(31), 3.

The magazine writers and the farmers. *Indiana Farmer, 25*(49) (1890), 7.

McMath, R. C., Jr. (1993). *American Populism: A social history, 1877–1898*. New York: Hill and Wang.

The new laws. (1891). *Indiana Farmer, 26*(11), 8.

Pencil. (1878). The lazy farmer. *Indiana Farmer, 13*(10), 7.

Phillips, C. J. (1968). *Indiana in transition: The emergence of an industrial commonwealth, 1880–1920*. Indianapolis, IN: Indiana Historical Bureau and Indiana Historical Society.

Politics in agricultural papers. (1884). *Indiana Farmer, 19*(29), 7.

Sneezer, Z. (1887). The conservative farmer. *Indiana Farmer, 22*(19), 14.

Stahl, J. M. (1884). The farmer and politics. *Indiana Farmer, 19*(34), 1.

State Alliance meeting at Ft. Wayne. (1890). *Indiana Farmer, 25*(26), 13.

State Farmers' Alliance organized. (1890). *Indiana Farmer, 25*(18), 8.

Trusler, M. (1888). Worthy Master's address. *Indiana State Grange 18th Annual Proceedings* (pp. 9–12) Indianapolis, IN: Indiana State Grange.

Washington County report. (1882). In *Annual Report of the Indiana State Board of Agriculture*. Indianapolis, IN: State Board of Agriculture.

Who sent them there? (1884). *Indiana Farmer, 19*(4), 9.

10

The Historian's Search for Civic Learning

This story of agricultural improvement in Indiana concluded with the agrarian populist movement of the 1890s, the rural drive for political and economic reform that prefigured crucial ingredients of the Progressive Era. The reader, therefore, might expect to be treated to some provocative reflections that raise questions about agricultural improvement's relationship to progressive reform, or, more broadly, about rural influences in the Progressive Era. One might begin by recalling a few things: the majority of the population at the twentieth century's turn lived in rural areas; a massive influx of farmers and ruralites poured into the cities; the upper Midwest was the hotbed of progressivism; and, the small town was the basis of progressive reformers' ideal community. Using these facts, the reader can formulate a speculative hypothesis and, then, check standard accounts of the Progressive Era to see if, perhaps, something is missing. There is a great deal of continuity between the 1850s and the early 1900s, particularly in civic aims and educational methods. Exploring the formative influences of progressivism, however, was not the purpose of this book. The mission was to reveal

Civic Learning through Agricultural Improvement, pages 221–240
Copyright © 2011 by Information Age Publishing
All rights of reproduction in any form reserved.

that civic learning took place among Indiana's farmers through agricultural improvement during the second half of the nineteenth century.

That point seems obvious. It is not and it is. If one thinks of civic learning as exclusively the outcome of civic education activities in high school classes, then one will not identify readily agricultural fairs and farmers' organizations as sources of civic learning. If, on the other hand, one takes a more comprehensive view of civic education, identifying its purpose (public good intentions) and effects upon those who engage in it (learning), then, it will seem obvious that agricultural fairs and farmers' organizations were prominent carriers of civic education in the rural communities of the nineteenth-century Midwest. Stated generally, the point is that civic education always has something to do with learning connected to community and the public good, but it need not take place in schools.

Given a moment to reflect, most people would agree that the preponderance of their civic learning—that is, what they learn about the behavior expected of good citizens, about their place in society, and about the demands the public (or government) can make legitimately upon their lives—is acquired not from schooling but by other means in their civic lives as adults. Among other sources, public policy, through what it prohibits and encourages, provides a constant form of civic education. Intermittent civic education and learning come from the media outlets and conversations that constitute public discourse about social conditions and policy alternatives. Belonging to various kinds of social institutions may orient a person to care about the broader community and people in it; taking an active role in social institutions may lead a person directly into public affairs and social action on behalf of particular causes. Episodic civic learning, but perhaps profound and far-reaching, may be acquired from major events in the life of a community that heighten awareness of normative expectations or the shared-ness of a common fate. Civic education of these sorts and others may be more or less systematic and structured, but people learn from it nonetheless.

A robust schooling experience is advantageous, clearly, for participating in and making sense of civic life. The objectives and learning activities of the school curriculum, however, do not map directly onto the idealized knowledge, dispositions, and behaviors possessed by citizens. Indeed, for various reasons, the school curriculum does not embody some things that are generally considered essential to the good of the civic community. In the late nineteenth century, improving agriculture ranked among the top priorities of Indiana's communities; yet, farm boys did not study agricultural science or economy in the common school curriculum. That fact says something significant about the relationship between schooling and society, about schooling's role within the complex of educational activities and institutions of nineteenth-century life.

In the historian's search for civic learning, the broad range of educational effort that orients people to ideas and behaviors sanctioned by the community (particularly its authoritative spokespersons) is of interest. Precedent for exploring education by means other than schools resides in the field of educational history's founding era (or rebirth) as a branch of historical inquiry in the late 1950s and early 1960s. At that time, many scholars embraced a robust conception of education broadly conceived, a conception that insisted that schooling be situated within a context of learning, as partly distinct from the social context of a time period. Equally important, scholars involved in the founding moment called on historians to investigate education in its fullness, to track the flow of educational endeavors from the sources of aims to policymaking and institution-building efforts to processes of learning (in and around the institutions' operations) to learning outcomes, and back again for the purpose of gauging the impact of education on the character and course of American society.

Two related landmarks are of particular significance in the founding era of the field of educational history, a report issued by the *Committee on the Role of Education in American History* (Buck et al., 1957) and Bernard Bailyn's lecture, *Education in the Forming of American Society* (1960). Of the two events, Bernard Bailyn's lecture is better known, for its damning indictment of educational history as it had been written by professional educators. Repeated numerous times over the years, the critique need not be revisited (although it and its companion essay are well worth reading). What is significant for present purposes is that Bernard Bailyn's interest in the education of history stemmed from his participation in the *Committee on the Role of Education in American History*. The Committee's report carried hints of Bernard Bailyn's disdain for educators' scholarship. It was also a frank admission by eminent historians, including Arthur M. Schlesinger, Sr. and Richard Hofstadter, that they found their own scholarship lacking (Cohen, 1976; Gaither, 2003).

The historians' problem was a methodological one. Historians had studied political affairs, literary developments, and social movements, but neglected their educational dimensions. They knew that educational efforts played a part in these things, but they knew relatively little about how those efforts intersected with prevailing tendencies. How, for example, did rural people in the late nineteenth century learn about the aims of the temperance movement? How did they (or not) become convinced that a temperate life was the life worth living? How was their receptivity (or indifference) to temperance conditioned by other concerns in their lives? To answer these kinds of questions, historians had little intellectual leverage. Without considering education broadly conceived as a dynamic force in people's lives, they could not "knit together the strands of American experience" (Buck et al., 1957, p. 3).

In what ways had deliberate efforts to reshape the habits and thoughts of others borne fruit (intended and otherwise)? If historians examined the actions of would-be educators and the responses of potential students, they could enrich (and, perhaps, revise) their understanding of how American society had transformed itself—to some extent, deliberately—over the generations. If learning could be said to have occurred, was it conceivable to hypothesize education as a driving force in history, akin to the frontier and industrialization?

Between its emphasis on education broadly conceived and its insistence that educational inquiries contribute to a broader understanding of American society, the *Committee on the Role of Education in American History* offered a stunning vision of education's possibilities for historical investigation. (Who, after all, places a hypothesis about education on par with Frederick Jackson Turner's frontier thesis?) Roughly two decades of intellectual excitement in the field of educational history followed. Only the initial impulse is attributable directly to the Committee's methodological challenge. It spurred people's thinking about intentions and impact. As a set of young scholars came of age during the turmoil of the 1960s, they discovered that schooling, as an institution of society, carried less-than-noble intentions. Schooling's impact in social life, most young scholars concluded, was not the kind that did justice to the ideals that Americans profess. The history of education broadly conceived could not compete with the history of schooling as a form of social criticism. The mission was largely abandoned.

Although occasional calls for renewed attention to education broadly conceived have been made, in recent years there seems to be little enthusiasm for the project (Tamura, 2010; Warren, 2003). The typical educational historian prefers to think of himself as an "educational researcher" whose mission is to build a body of scholarship about schooling (Rury, 2006, p. 597). The "educational researcher" designation claims far more than its school-minded proponents care to deliver; dropping the word, history, from the label signifies a rejection of the ideals of historicism. As a work of educational history, therefore, this case study of civic learning through agricultural improvement represents a recovery effort. In more ways than one, it sought to reach into the past to reclaim crucial elements that recent historians have neglected. Above all, though, it sought to recapture something of the spirit that animated the *Committee on the Role of Education in American Society.*

Guidance came from two sources, principally. The first was a "Memorandum" that Richard Storr (1976) wrote for the Committee as a product of its discussions. The second was Richard Storr's (1961) account of how he put into action the mandate to seek the workings of education broadly conceived, an essay titled, "The Education of History: Some Impressions." One might think of the Memorandum as a conceptual piece for identifying education and perceiving the fullness of its processes in historical settings. The Impressions essay is best characterized as a guide for doing educational history, for looking for learning in whatever guise it appears and making sense of it. Both documents point directly toward fulfilling the Committee's mission of using the study of education to enrich understanding of broad transformations in American life.

The reader is invited to give Richard Storr's essays serious consideration, as well as the reports of the *Committee on the Role of Education in American Society*; an in-depth analysis exceeds present purposes. Instead, five key ideas expressed within them will be discussed and related to this case study of civic learning through agricultural improvement. The ideas are: (a) the selection of a topic; (b) the selection of an institution; (c) a cautionary warning about teleology in reverse; (d) institutional operations and policymaking as sources of civic learning; and, (e) placing education in the mainstream of American life. Discussion of these ideas will be somewhat personal, but the hope is to provide insights to historians who might consider looking for learning in whatever guise it appears, particularly civic learning.

Topic Selection

It seems axiomatic that a historian's research questions ought to be derived, principally, from the life of the past. One need only browse the library catalogue to gain an impression of how far recent historians have deviated from this principle. In the nineteenth century, the vast majority of Americans lived on farms; yet, few recently published books on agriculture will be found; rural communities and social life fare somewhat better. If one examines historical textbooks written for high school and collegiate survey courses, one finds that agriculture is strangely neglected and that, by the second half of the nineteenth century, the terms, agriculture and rural, seem to apply only to the South and West. The field of educational history, in particular, seems to be only vaguely aware of just how rural the United States was in the nineteenth century. Devoting a rather disproportionate amount of attention to interpreting urban schooling, educational historians have largely neglected rural schooling (outside the Deep South). They have displayed little awareness of the educative efforts that took place outside formal schooling, but that were instrumental responses to the conditions of life for people who resided in rural communities throughout the United States.

Grappling with the question of how educational efforts reshaped the character of society over time, demands that the starting point for inquiry be grounded in the concerns voiced by past generations. Declining soil fertility, diversifying local economies, and strained relations between town and country are not our concerns. They were concerns of nineteenth-century rural communities that received a variety of educational responses. Failure to attend to the concerns of the past has caused most historians not to notice.

Guilty, as charged. The goal of finding civic learning among farmers was not the initial impulse behind this case study. Oddly, it seems in retrospect, that the initial mission was to explore how scientific knowledge was diffused among the common people of the nineteenth century. Knowing that most people were farmers established the baseline, and encountering the experimental results of one farmer's experience growing sweet sorghum in the 1850s, pointed toward county agricultural societies and fairs as diffusion mechanisms. With a little more foraging in the documentary record, it became apparent that Indiana's farmers were not very receptive to scientific agriculture (diversified commercial farming) in the 1850s.

In light of scientific agriculture's profitability (as demonstrated by its advocates), farmers' indifference to it raised a question: What made improvement-minded farmers different from other farmers? A second question followed: How did agricultural improvement's educational institutions attempt to close the gap between the farmers? Their difficulties and repeated efforts raised a third question: What was so important about agricultural improvement that its advocates would try and try again with innovations and new institutions? Much of the answer pointed away from the farm and toward the town, and it was only partly related to scientific agriculture. Going with the flow set by the primary sources yielded a steady diet of surprises and new questions that called for answers, along with an entirely different research project.

Undeniably, one might launch a historical investigation with a predetermined topic that is inspired by phenomena of the present and stick to it. Pursuing the topic (whatever its origins), however, requires delving deeply into the particulars. On first impression, the scientific agriculture of the 1850s does not resemble modern science. Much of it was moralizing supported by itemized statements of expenditures and receipts. Something similar can be said about historical civism. Making sense of how a different era—and different people within it—thought of community and obligations to the public good demands sustained engagement in the life of the past.

What appears to be mere "nuances of coloration" may be significant differences and departures. To grasp education as part and parcel of people's lives requires exercising "a very keen sense of the concrete, a sense that bears the same relation to an antiquarian love of particulars that uranium does to lead" (Storr, 1961, p. 125). If one held the view that a farmer is a farmer is a farmer, little of the turmoil within the Grange over partisan politics and economic cooperation would get noticed. (One might conclude simply "The farmers decided..." and leave it at that.) Likewise, alterations in county fair premiums could be overlooked easily, even though the allocation of prize money mattered a great deal. In the 1870s, the fair that bestowed large sums upon fast horses and professionals with show herds was not considered by farmers to be a real agricultural fair. Yet, in the 1880s, a real agricultural fair could have horse races and fancy cattle, supplied by local farmers. The marked decrease in premium award sums signaled something significant about broader changes in the agricultural fair.

Institution Selection

At the risk of flogging the horse to death, the starting point for inquiry into the education of history is not the question, what did schools do, how, and why? This rule applies generally, except for those unusual cases in which schooling happened to be the only institution used to carry out a specific ambition. Just how one might know what are the unusual cases without examining institutions other than schooling is a mystery worth contemplating. The generic baseline questions for inquiry run something like this: In response to changes in society (real or imagined), what needs for learning did some people perceive as unmet? What actions did those people take in the hope of meeting those needs? How did other people react, particularly the intended learners? What impact did educational efforts—and the events set into motion by them—have on people's sensibilities? To hypothesize answers to these questions, the historian browses the historical landscape. He seeks prominent concerns voiced in the primary sources and tracks down whatever emanates from them. The quest is for something that "can sensibly be described as educational" in light of the contours of the age (Storr, 1961, p. 124).

The agencies created for farmers' education (agricultural associations, fairs, and institutes) had peculiar purposes, but they were not at all unusual. They were amalgamated derivations from a wide range of educational institutions, from scientific societies and lyceums to working men's associations, fraternal organizations and literary societies to market fairs and public exhibitions. The range of educational institutions in nineteenth-century life speaks to the variety of uses to which education was put. It also suggests that overlapping purposes infused specific educational institutions. An institution might appear to be specialized or single-purpose; the special purposes of familiar institutions, however, appear in the most unlikely of places. A county fair is an odd place to find education in standards of beauty. With oil paintings, line drawings, lithographs, and flowers stuffed into Floral Hall, one should not rule out the possibility that some people did learn something about good taste at the county fair.

The multipurpose nature of institutions (designated educational or otherwise) is particularly applicable to institutions that operated in rural communities. The countryside was not quite as isolated as some people claimed, but the barriers to travel were considerable. When people gathered, the official purpose of an institution shared space with other priorities. Within the Grange, for example, efforts to put politics and public policy outside the gates never quite succeeded; ejecting economic cooperation almost destroyed the institution in Indiana. With its subordinate branches pared down, primarily, to a combination social club and agricultural society, the Grange's mission was more manageable. The problem was not simply that common farmers wanted the Grange to do things that prominent farmers did not want it do for civic reasons. Farmers brought far more into the Grange than any single institution could withstand. Fewer members and fewer purposes was the price of institutional survival.

Multiplied across the landscape, experiences of this sort made some people aware of the value of single-purpose institutions. They learned to prefer a focused mission and to build in safeguards against mission creep. The rural nature of American society made most people slow learners with regard to this lesson. In the Progressive Era, although institutional lines were defined more clearly, overlapping purposes and multiple burdens characterized most institutions that were embedded in community life. The schoolhouse that served as a social center for adults on appointed evenings and holiday celebrations, as a de facto town hall, and as an occasional dispensary site for medical supplies and advice was no aberration. In light of these considerations, in the search for the education of history the baseline for institutional selection should be to assume that unlikely institutional candidates are carrying out the educational mission being pursued, however imperfectly.

What of the activities carried out in institutions that appear to possess educational purposes? How does one identify something that can sensibly be described as educational? Triangulation—using what is known about the targeted audience's learning in other circumstances as intellectual leverage—is essential. There are historical accounts of agricultural improvement that assume that common farmers were active members of agricultural societies in the early nineteenth century. That the assumption is ill-founded is corroborated readily by weighing the level of educational attainment among farmers against the polished papers delivered at the meetings. Agricultural improvement's formal curriculum of essays and papers was not farmers' principal source of agricultural learning. The typical farmer did not learn that way. Nor did he subscribe to the agricultural papers that reprinted literary addresses.

Most farmers were functionally literate, but they were unaccustomed to doing extensive reading. They learned best from experience, by doing and observing. They learned most of their agricultural knowledge from working the farm as youths, under parental tutelage (hence the frequent complaint that farmers simply followed the beaten tracks of their fathers). Farmers learned some agricultural knowledge from visiting other farms and trial-and-error experiences on their own farms. The displays, demonstrations, and conversations that took place at genuine agricultural fairs concentrated these kinds of learning and heightened farmers' perception. Placed against farmers' typical learning experiences, the serious events of the agricultural fair can be classified sensibly as educational events. The degenerate events improving farmers deplored were also learning episodes, but not for agricultural learning.

Even at its agricultural best, the fair was an unreliable agricultural educator. Too much was left to chance and too little of the why and how of agricultural techniques and products was actually explained. Complaints about farmers' erroneous applications of agricultural improvement were exceeded only by complaints about their indifference to it. Understandably, some people surmised that the errors and indifference were related. The Grange meeting's focused papers on practical agriculture and follow-up discussions were much more reliable than the fair. With the Grange's decline, they reached only a few thousand of Indiana's farmers. The farmer institute's program combined the fair's demonstration feature and the Grange's discussion feature into a fairly effective package of agricultural education. Prior to that, there was considerable slippage between the agricultural knowledge presented in formal papers and how most farmers applied it. Institutions devoted to agricultural education were poor conveyors of agricultural knowledge. They were far more potent as sources of civic learning.

Teleology in Reverse

With the phrase "teleology in reverse," Richard Storr referred "to the tendency to read explicit intention back into the past, to think that innovators had planned from the beginning to accomplish what we see as their ultimate achievement" (Storr, 1976, p. 334). Two things are at issue in teleology in reverse: the investigator's sense of an educational institution's original purpose and his awareness of the outcome, its enduring shape and mission. Equipped with hindsight, might not the historian peer into an institution's founding moment with a narrow gaze? Would he be inclined to perceive clarity of vision and purpose in changes made to the institution?

What might he overlook? He might fail to notice that, at the institution's moment of creation, its purposes were not clearly formulated by its founders. He might also fail to take seriously claims made by other people as to what the institution ought to do. As a consequence, he might not perceive some of the action that took place in and around the institution's operations and policies, pulling it back and forth between competing demands. Multiple purposes, constituencies, and disputes might have been involved, but a tidy pattern of development appears nonetheless. (The flipside of teleology in reverse is too much irony, but this seems to be less of a problem in accounts of educational institutions.)

The corrective to teleology in reverse is not, as in constitutional law, to latch onto the definitive original intent of an institution's principal founders. Characteristically, that approach is used to deny legitimacy to rival ideas. The jurist's mission is not the historian's. To the historian, founding moments are mushy, not solid, and the developments that flow from them are usually tortuous, stocked with indecision and reversals of policy. The action, not the initial intent or the outcome, grabs the historian's attention. When lost in a forest of conflicting aims, competing priorities, and quarreling constituencies the historian has reached his destination. The task of narrative creation leads him home.

Identifying a single original intent in agricultural improvement proved impossible. To the men involved in the 1850s, the strictly agricultural and the civic agenda were inseparable. It did not matter whether farmers became good farmers or good citizens because the two labels covered the same set of behaviors and dispositions. Likewise, it did not matter whether the infrastructure was improved first, or if farmers began to intensify their agricultural practices and then demanded infrastructural improvements.

The early agricultural societies were staffed with leading men who wanted both things to occur simultaneously. Both were ends of themselves; each stimulated the other.

The general run of farmers, however, showed little interest in county agricultural societies and agricultural improvement as it was conceived by leading men. One can only imagine what the farmers thought were the real purposes of the agricultural society and fair (and surmise that they were likely to be things that their sponsors left unspoken). The agricultural society's failure to capture farmers' fancy set the stage for a divergence of aims. Serious-minded farmers wanted a strictly agricultural fair; gentlemen farmers wanted a public forum to show their worth; and, leading townsmen wanted the fair to be a civic (town-boosting) institution. Were the rival aims immanent in the founding moment? Did merchants, in particular, always want the agricultural fair to fill their pocketbooks and to jumpstart local economic growth with infusions of capital from abroad? A civic booster organ, after all, was what the agricultural fair became rather quickly.

Based upon the pattern of events and surviving documents, answers to these questions point toward indeterminacy in the emergence of purposes, not toward the identification of single-minded intentions that became realized over time. In the early 1850s, leading farmers and townsmen (who delivered most addresses at meetings and fairs) thought that common farmers' indifference to self-improvement was the overriding obstacle to their communities' improvement. Agricultural addresses focused on the benefits to the farmer of improved agriculture, infrastructure, and access to urban markets. They did not mention the benefits that hosting fairs might confer upon towns. Yet, one must wonder: were not town merchants and businessmen aware that fair-hosting would be a boon to local commerce? To some degree, surely, townsmen must have been alert to the agricultural fair's pecuniary potential.

The earliest agricultural fairs were not big money-generators though. The wave of railroad building that swept Indiana in the late 1850s gave agricultural fairs their economic potential. Without the railroad, the county seat's merchants lacked a superabundance of goods, and there was little hope of attracting capital investment from abroad. With the railroad, selling goods to farmers and promoting economic development through public relations campaigns became real possibilities. A change in environmental conditions reshaped intentions: serious farmers were right to insist that the agricultural fair had been perverted from its original design, even though the design had never been strictly agricultural.

The railroad building of the late 1850s reworked how leading towns-men thought about the relationship between their towns' economic development and the imperative to improve farmers' human capital. If local economic development could be fostered independently of farmers, why wait for farmers to catch the spirit of improvement? By that time, since farmers failed to participate in agricultural societies, leading townsmen held the fair and the reins. It was always intended that the agricultural fair should be the county fair, but the county fair that was taking shape at the end of the 1850s was not what people had in mind when the organic act for agricultural improvement was passed in 1851. In its founding moment—while agricultural fairs were still a novelty in many of Indiana's counties—changes were made to the fair as a result of what people in rural communities were learning about each other, the fair's capacities, and society at large.

Institutional Operations and Policymaking as Sources of Civic Learning

"Institutions serve to bring ideas to bear upon men, and to strengthen (or to weaken) the capacity to deal with ideas" (Storr, 1976, p. 341). Educational agencies are created to change people's minds. It is also the case that newly created educational agencies, as well as innovations to existing agencies are products of learning. Their founders have discerned something in the environment around them and concluded that it warrants revision to the existing stock of educational supply. The solutions they devise are, to varying degrees, extensions of their personal experiences and sensibilities. As innovators, the men who hatch new educational ideas are, by definition, unusual men; either they have had unusual experiences or they have acquired particularly acute perception.

Necessity, and at least a touch of the reformer's zeal, takes unusual men and their strange ideas into society. The challenge is to persuade other people—who have had somewhat different experiences and possess somewhat different ideas—of two things: that the innovators' notions about what is happening in society (or will happen) are not products of delusion; and, that the proposed educational solutions are well calculated to produce desirable effects. The educational idea must command moral legitimacy and its institution material resources; public support must be generated and policies fashioned (Storr, 1976).

Launching an educational idea into the currents of society marks the beginning of a "process of spinning a web of policy [that] goes on continuously" (Buck et al., 1957, p. 8). Guiding policies are devised; an educational institution is created; operational policies are devised; people's encounters with the institution spark calls for revision to its policies and operations. Does the institution function properly to meet its stipulated aims? Are the stipulated aims the proper ones? What do targeted students actually learn from their experiences with the institution? As people scrutinize the institution, debate its merits and faults, and exchange ideas and experiences in a recursive process spread over time, "a fabric of decisions" unfolds. It is in the policymaking life in and around an educational institution "that the direction of educational effort is fixed and that its efficacy is determined" (Buck et al., 1957, p. 8). The educational institution's impact upon targeted students may be the object of attention, but a great deal of learning may take place among the people at large who are interested in the students' and the institution's fate. Some learning may be about education, that is, it may concern curricula, instruction, and their supporting requirements. Much of the learning may be civic learning.

To the extent that the institution's operations and policy life draw people of unlike minds and experiences into a shared, if disputed, concern, a form of civic learning can be said to be occurring. The learning is not civic education in its totality (in the modern democratic sense), for the participants in an institution's policy life may not represent the entire community, and the institution itself may not be designed to serve directly the general public. Nevertheless, people with different life conditions and priorities are compelled to consider each other and rival claims about the good. From their interaction in policy debates, they learn something about the nature of their community. From the policy outcomes and the manner in which decisions are made, they learn something about their proper place in community. Farmers learned something about leading townsmen's esteem for agriculture when they took advantage of lightly attended agricultural meetings to turn the county fair into a horse racing and gambling festival.

Often people do not like what they learn; what they learn may destroy the shared venture. Sometimes, though, people's stake in the common institution is sufficiently strong to keep them in the fold. As long as they stay engaged—even if their intent is to overturn established policy and practice—there is a chance that civic learning will continue. The aggrieved may change other people's minds. They may change their own minds with additional doses of reflection and experience. They may become resigned to the unchangeable. A new steady state of consensus on how things ought to be takes shape. A gap between unlike minds closes. The closed gap may or may not be a robust (or appealing) victory for civism, but the essence of civic learning has occurred. Leading grangers knew this or something like it. It was part of the reason why they refused to permit farmers an exit option from the county fair. They checked the township fair movement and insisted that farmers mobilize to reform the county fair.

This kind of educational process can be cantankerous and volatile. That potential makes it a potent source of civic learning. Institutional operations and policy making were the driving forces in the civic learning generated by agricultural improvement's agencies. The potential for civic learning resided in the social creativity that flourishes in verges of rival ideals, contrary conventions, and competing priorities (Boorstin, 1989, p. xv). There were many verges in the terrain covered by agricultural improvement: where the drawing power of distant cash crop markets met up with prescriptions for mixed husbandry in farmers' decision-making; where serious agriculture met up with entertainment at the county fair; where true progress met up with town-boosting mania in economic development; where the farmer's poorly conceived self-interest met up with civic concerns in the Grange; where the State Fair's debt burden met up with demands for farmer institutes in the State Board of Agriculture. Agricultural improvement did not create these rival ideals and competing priorities, but its educational efforts created spaces for the verges to form. Farmers did not learn much agricultural knowledge from their encounters with others in these verges, but they did learn about other farmers and other members of their civic communities.

Individuals' civic learning in agricultural improvement's verges in Indiana must be left largely to the imagination. Few encounters were recorded by private individuals and institutional records are too sanitized and spare of detail (even the General Assembly failed to record debates). What did the common farmer think when the State Grange leadership killed the State Purchasing Agency, or when the prominent granger refused him credit at the cooperative store? What did the town banker think when a crowd of farmers brought their discontent with the agricultural fair's depravity into the usually empty meeting room? What, ultimately, convinced a leading townsman to lend his mite to farmer institutes? Was it the accumulation of bits of learning? Or, was it simply the fact that other townsmen were doing it and were urging him to fall in line? What of the farmer who was dependent on a single cash crop for his family's livelihood? Did he learn that diversified agriculture was in his self-interest only, or did he learn to think of himself as a public benefactor of the local community? We cannot declare on the particular case. We can infer from the aggregate, from the changes that were made to the institutions of agricultural education. Such changes were not direct reflections of individuals' civic learning. They do reveal shifts in public opinion, though, and these shifts can be classified reasonably as outcomes of civic learning.

A point should be stated directly about farmers' civic learning from the policy life in and around their agricultural educational agencies: policymaking processes did yield, ultimately, the civic outcome desired in the 1850s, but the ways in which farmers learned to be good farmer-citizens were not how agricultural improvers wanted farmers to learn. Ideally, farmers would have joined the county agricultural society, engaged in book-farming, and acquired a ken that encompassed economic development policies and the fortunes of the county seat. Exposure to agricultural education was supposed to guide farmers smoothly toward a better state of local citizenship. That, clearly, did not happen.

Agricultural education institutions were intended to lead common farmers toward the kinds of behavior and values advocated by prominent farmers. A qualifier must be affixed: when enacted and received properly by sponsors, instructors (leading farmers), and students (other farmers). Agricultural improvement's education was seldom done properly. In the 1850s, agricultural society meetings were too serious to appeal to practical farmers. Townsmen soon emptied the agricultural society and fair largely of agricultural content, and offered farmers horse races. Improvement-minded farmers stayed home.

Farmers joined the Grange in droves, only to be alienated by the leadership's refusal to support their political and economic reform goals. After the farmers' movement receded, and relatively few farmers remained in it, the Grange functioned more as a social club than as a voluntary association devoted to agricultural education. The industrial associations' members did not need agricultural instruction for themselves and they did almost nothing to offer it to other farmers. Discontent with existing institutions prompted improvement-minded farmers to try yet another in the late 1880s. The farmer institutes functioned, more or less, as they were designed, although a sizable contingent of farmers did not attend.

It took two generations of trial and error, reversals, and restarts before the educational intentions and expectations of sponsors, teachers, and students matched. A more dysfunctional case of educational effort, of conflicting aims and faulty instructional design, can hardly be imagined (although it may be closer to the norm than we might expect). Yet, somehow, between the 1850s and the 1880s, the majority of Indiana's farmers came to agree, more or less, with leading farmers and townsmen on the desired aims for the rural community and their role in it. Can the sequence of failed efforts, perverted intentions, and mismatches between sponsors, teachers and students in agricultural education be considered a civic learning process? Can the convergence of public opinion that ultimately occurred in the late

1880s be considered an outcome of civic learning? If so, the policy-spinning actions of common farmers, leading farmers, and townsmen who brought their ambitions and frustrations to bear upon each other through their shared agricultural institutions must be considered influential sources of civic learning.

Placing Education in the Mainstream of American Life

"To what extent is the American the fruit of happenstance and to what extent is he the product of design?" (Storr, 1976, p. 335). That pregnant question crystallized the mission of the *Committee on the Role of Education in American History*. The Committee was interested in "education as a creative force in American history," in reconstructing how education contributed to the broader currents of change (Buck et al., 1957, p. 6). It was and is chimerical to attempt to isolate a sphere of education from the contours of social life in pursuit of hypothesized results that can be attributed exclusively to educational effort. Yet, it is possible to envision education operating as a help or hindrance to the fulfillment of other priorities, as an ingredient in the remaking of American character over the generations.

It is also possible to formulate an overarching hypothesis about education's changing role over time. The Committee did this, hypothesizing that, in response to accumulating social crises, Americans have relied increasingly on formal educational institutions for the young to mediate change, whether to check or advance it, or simply to adapt to what is perceived as inevitable. Something resembling this hypothesis has become conventional wisdom. It is worth continued scrutiny in its application to different eras, places, and issues. If we weigh efforts to use formal schooling against other educational efforts, as well as against non-educational policies that were directed toward similar ends, we might find that schooling was not the solution preferred by most Americans. (While common school reformers championed free schools common farmers demanded free land.) We might find that an innovation in educational policy was not the most important solution but was, rather, an auxiliary force, the third-best alternative, or the only measure put into action that was practicable politically. Whatever the hypothetical case might be, recognizing it requires placing education in the mix of forces shaping the life of society. The Committee's report was an invitation for historians to do that.

In its report, the Committee identified eight broad trends in American history in which they thought educational efforts may have played a decisive role. Three of those themes have a direct bearing upon this case study of civic learning through agricultural improvement in nineteenth century

Indiana. The first is the utilization of material resources through efforts to develop human capital; the second is the role of educational institutions in the building of new communities on the frontier. As they apply to the case study, these two themes must be considered as subsets of a third, the transformation of American society from a rural and agricultural society to an urban and industrial society. As the Committee acknowledged, this involved a "great transformation in attitudes" toward social institutions and relationships with other people. At its heart lay "the displacement of the older 'individualism' by 'socialization, '" of men's determination to keep society's demands at arm's length by men's realization that their families' lives were inescapably bound up with that of their surrounding communities (Buck et al., 1957, pp. 12–13). The Committee's concern was similar to what ultimately became the governing question of this case study: How did Indiana's farmers learn to become civic-minded members of their local communities?

In the five decades since the *Committee on the Role of Education in American History*, historical scholarship has pursued its theme of rural and urban society, individualism and community, in various ways. For explaining the early nineteenth century, a powerful hypothesis about economic change now stands alongside the consequences of the American Revolution as a pivotal force remaking the character of early American society. Summarized, the hypothesis is that a market revolution occurred, that a great transformation in values, behavior, and social institutions accompanied the growth of commercial production and exchange in the early United States. Farmers began producing for sale to distant markets, internal improvements and railroad building captured public attention, and non-farming industries grew rapidly. Americans embraced economic progress. They became more individualistic and less oriented to traditional norms governing personal behavior and obligations to extended families. They participated in a flourishing print culture that fed them a steady diet of practicality, morality, and sentimentality that were, in many ways, at variance with the teachings of previous generations. A distinctive town-dwelling life crystallized. By the 1830s, the market revolution was all but complete (Sellers, 1991; Wood, 1991).

The market revolution hypothesis has great explanatory power and it is very convincing. However, there was an exception to it: the vast majority of the population. Farmers became participants in the market economy through seasonal sale of cash crops, but their lives were not transformed by it immediately. Outside the vicinity of sizable towns, farmers' contact with market exchange was limited; most appeared to be in no hurry to increase it. Indeed, some scholars have pointed to a considerable body of evidence that suggests that farmers opposed efforts by merchants and businessmen to make their communities fully a part of the commercial economy through infrastructure projects and public policy. Farmers' opposition was grounded in attachments to traditional values such as familial independence, rough equality, and cooperation among neighbors. The market economy's reach into their lives (supported by increasing population pressure on the land), however, meant that commercial farming could not be avoided (Clark, 1990; Henretta, 1992; Kullikoff, 1992).

As more farmers engaged in commercial farming, one might say, they did so in pursuit of traditional goals: preserving (as much as possible) familial independence and their capacity to pass on farmland to their sons. Despite the allurements of the emerging middle class town-dwelling life, farm families remained aloof. Careful scholars of the market revolution—including its greatest proponent, Charles Sellers—took note of this, but it has not penetrated historical consciousness. The limitations of the market revolution hypothesis, therefore, raise the question: Assuming that the market economy does not operate in people's lives using an invisible hand, how did farmers learn to want to become participants in the emerging town-centered and market-infused civic order by the late stages of the nineteenth century?

Opposite the market revolution, at the close of the nineteenth century, the prevailing hypothesis is that rapid industrialization and urbanization induced a search for order between the upper and middle classes of society (Wiebe, 1967). In the hands of professionals and white-collar managers who were linked together through nationwide corporations, institutions, and print culture, progressive reform attempted not only to temper the excesses of industrialization, but to infuse rationality and a peculiar set of values and behaviors into social institutions and the private reaches of social life. The result in the rural areas of the northern United States, as historian Hal Barron (1997) reveals, was a mixed harvest. Resistant to outside and urban pressures, the countryside blocked and frustrated key elements of progressive reform, particularly school consolidation and public road administration by county and state agencies. Yet, much of the progressive package—symbolized by the Farm Bureau, cooperative dairies and grain elevator associations, mail-order catalogues, and the Model T—was embraced by many farm families and rural inhabitants. One might wonder: What prepared the way for this complex and differentiated (dare I suggest, nuanced?) rural response to progressivism?

There is a great lacuna in our historical understanding of what took place in American society between the market revolution and the organizational synthesis, between the beginning and the end of the nineteenth century. Perceptively, Hal Barron did not locate the beginning of the "second great transformation" of rural life in the 1890s (Barron, 1997, p. 8). He marked 1870 as the starting date. Without discounting in the least Hal Barron's scholarship, he might have selected the 1850s, instead. At that time, in the Old Northwest (the center of gravity for the rural North), the imperatives of the market revolution were taken for granted among townsmen. But the preferred means for advancing progress—state-sponsored internal improvements—had been set back by a farmer's backlash. Midwestern farm families were still seasonal and limited market participants; as participants in community, they were still guided largely by the traditional ethos of rural separatism. In the 1850s, town-dwellers regained the initiative and found other means to bring the imperatives of the town-dwelling life to bear upon farmers. Commitments to their guiding ideals—as well as the mechanisms for carrying them out—were in place before the Civil War. Ambitions intensified as a result of the war's impact, and the mechanisms for carrying out economic development became permanent fixtures of community life in the 1870s.

One of the great means townsmen and their leading farmer-allies latched onto in the 1850s was agricultural improvement. This case study has attempted to reveal some of what and how farmers learned from it. What Indiana's farmers learned was not exclusively centered upon their roles as producers and consumers in the market economy; nor was it exclusively concerned with their participation in formalized institutions that carried the imperatives of the town-dwelling life. Rather, it fell somewhere in between. Most farmers had not learned the lessons of the market revolution by the 1850s. They had not learned to maximize their farms' profit-making potential in ways that might have fostered a harmony of interests between town and country. To teach them, institutions were created, but town-dwellers (and farmers) were slow to learn the lessons about society that progressive reformers of the 1890s took to heart in their search for order. If this case study of civic learning through agricultural improvement has succeeded, it has offered insights into the learning processes that bridged these two great hypotheses about the currents of change in nineteenth-century life.

This case study has offered only a partial answer to the civic question, how did Indiana's farmers learn to become civic-minded members of their local communities? From it, however, the reader may be convinced that Indiana's farmers did, in fact, learn something about their civic obligations from their encounters with agricultural improvement's educational institutions. The reader may also be convinced that the history of education broadly conceived is worth pursuing. If little else, this case study may serve a cautionary warning: we may know less about the civic education of history than we think.

References

Bailyn, B. (1960). *Education in the forming of American society: Needs and opportunities for study.* New York: W. W. Norton.

Barron, H. S. (1997). *Mixed harvest: The second great transformation in the rural north, 1870–1930.* Chapel Hill: University of North Carolina Press.

Boorstin, D. J. (1989). *Hidden history: Exploring our secret past.* New York: Vintage Books.

Buck, P., Faust, C. H., Hofstadter, R., Schlesinger, A. M., & Storr, R. J. (1957). *The role of education in American history.* New York: Fund for the Advancement of Education.

Clark, C. (1990). *The roots of rural capitalism: Western Massachusetts, 1780–1860.* Ithaca, NY: Cornell University Press.

Cohen, S. (1976). The history of the history of American education, 1900–1976: The uses of the past. *Harvard Educational Review, 46*(3), 298–330.

Gaither, M. (2003). *American educational history revisited: A critique of progress.* New York: Teachers College Press.

Henretta, J. (1992). *The origins of American capitalism.* Boston: Northeastern University Press.

Kullikoff, A. (1992). *The agrarian origins of American capitalism.* Charlottesville: University of Virginia Press.

Rury, J. L. (2006). The curious status of the history of education: A parallel perspective. *History of Education Quarterly, 46*(4), 571–598.

Sellers, C. (1991). *The market revolution: Jacksonian America, 1815–1846.* New York: Oxford University Press.

Storr, R. J. (1976). The role of education in American history: A memorandum for the committee. *Harvard Educational Review, 46*(3), 331–354.

Storr, R. J. (1961). The education of history: Some impressions. *Harvard Educational Review, 31*(2), 124–135.

Tamura, E. H. (2010). Value messages collide with reality: Joseph Kurihara and the power of informal education. *History of Education Quarterly, 50*(1), 1–33.

Warren, D. (2003). Beginnings again: Looking for education in American histories. *History of Education Quarterly, 43*(3), 393–409.

Wiebe, R. (1967). *The search for order, 1877–1920.* New York: Hill and Wang.

Wood, G. S. (1991). *The radicalism of the American revolution.* New York: Vintage Books.

LaVergne, TN USA
28 December 2010
210239LV00001B/34/P